Applied and Numerical Harmonic Analysis

Series Editor
John J. Benedetto
University of Maryland

Editorial Advisory Board

Myoung An
Andrzej K. Brodzik
Richard Tolimieri

Ideal Sequence Design in Time-Frequency Space

Applications to Radar, Sonar, and Communication Systems

Birkhäuser
Boston • Basel • Berlin

Myoung An
deciBel Research, Inc.
1525 Perimeter Parkway, Suite 500
Huntsville, AL 35806, USA
myoung@dbresearch.net

Andrzej K. Brodzik
The MITRE Corporation
202 Burlington Road, MS E050
Bedford, MA 01730, USA
abrodzik@mitre.org

Richard Tolimieri
Psypher, Inc.
1327 Monte Sano Blvd. SE
Huntsville, AL 35801, USA
richard@psypher.org

ISBN 978-0-8176-4737-7   e-ISBN 978-0-8176-4738-4
DOI 10.1007/978-0-8176-4738-4

Library of Congress Control Number: 2008932706

Mathematics Subject Classification (2000): 42-04, 94A99, 54A99

Printed on acid-free paper

www.birkhauser.com

ANHA Series Preface

The *Applied and Numerical Harmonic Analysis (ANHA)* book series aims to provide the engineering, mathematical, and scientific communities with significant developments in harmonic analysis, ranging from abstract harmonic analysis to basic applications. The title of the series reflects the importance of applications and numerical implementation, but richness and relevance of applications and implementation depend fundamentally on the structure and depth of theoretical underpinnings. Thus, from our point of view, the interleaving of theory and applications and their creative symbiotic evolution is axiomatic.

Harmonic analysis is a wellspring of ideas and applicability that has flourished, developed, and deepened over time within many disciplines and by means of creative cross-fertilization with diverse areas. The intricate and fundamental relationship between harmonic analysis and fields such as signal processing, partial differential equations (PDEs), and image processing is reflected in our state-of-the-art *ANHA* series.

Our vision of modern harmonic analysis includes mathematical areas such as wavelet theory, Banach algebras, classical Fourier analysis, time-frequency analysis, and fractal geometry, as well as the diverse topics that impinge on them.

For example, wavelet theory can be considered an appropriate tool to deal with some basic problems in digital signal processing, speech and image processing, geophysics, pattern recognition, biomedical engineering, and turbulence. These areas implement the latest technology from sampling methods on surfaces to fast algorithms and computer vision methods. The underlying mathematics of wavelet theory depends not only on classical Fourier analysis, but also on ideas from abstract harmonic analysis, including von Neumann algebras and the affine group. This leads to a study of the Heisenberg group and its relationship to Gabor systems, and of the metaplectic group for a meaningful interaction of signal decomposition methods. The unifying influence of wavelet theory in the aforementioned topics illustrates the justification for providing a means for centralizing and disseminating information from the broader, but still focused, area of harmonic analysis. This will be a key role of

ANHA. We intend to publish with the scope and interaction that such a host of issues demands.

Along with our commitment to publish mathematically significant works at the frontiers of harmonic analysis, we have a comparably strong commitment to publish major advances in the following applicable topics in which harmonic analysis plays a substantial role:

Antenna theory	*Prediction theory*
Biomedical signal processing	*Radar applications*
Digital signal processing	*Sampling theory*
Fast algorithms	*Spectral estimation*
Gabor theory and applications	*Speech processing*
Image processing	*Time-frequency and*
Numerical partial differential equations	*time-scale analysis*
	Wavelet theory

The above point of view for the *ANHA* book series is inspired by the history of Fourier analysis itself, whose tentacles reach into so many fields.

In the last two centuries Fourier analysis has had a major impact on the development of mathematics, on the understanding of many engineering and scientific phenomena, and on the solution of some of the most important problems in mathematics and the sciences. Historically, Fourier series were developed in the analysis of some of the classical PDEs of mathematical physics; these series were used to solve such equations. In order to understand Fourier series and the kinds of solutions they could represent, some of the most basic notions of analysis were defined, e.g., the concept of "function." Since the coefficients of Fourier series are integrals, it is no surprise that Riemann integrals were conceived to deal with uniqueness properties of trigonometric series. Cantor's set theory was also developed because of such uniqueness questions.

A basic problem in Fourier analysis is to show how complicated phenomena, such as sound waves, can be described in terms of elementary harmonics. There are two aspects of this problem: first, to find, or even define properly, the harmonics or spectrum of a given phenomenon, e.g., the spectroscopy problem in optics; second, to determine which phenomena can be constructed from given classes of harmonics, as done, for example, by the mechanical synthesizers in tidal analysis.

Fourier analysis is also the natural setting for many other problems in engineering, mathematics, and the sciences. For example, Wiener's Tauberian theorem in Fourier analysis not only characterizes the behavior of the prime numbers, but also provides the proper notion of spectrum for phenomena such as white light; this latter process leads to the Fourier analysis associated with correlation functions in filtering and prediction problems, and these problems, in turn, deal naturally with Hardy spaces in the theory of complex variables.

Nowadays, some of the theory of PDEs has given way to the study of Fourier integral operators. Problems in antenna theory are studied in terms of unimodular trigonometric polynomials. Applications of Fourier analysis abound in signal processing, whether with the fast Fourier transform (FFT), or filter design, or the adaptive modeling inherent in time-frequency-scale methods such as wavelet theory. The coherent states of mathematical physics are translated and modulated Fourier transforms, and these are used, in conjunction with the uncertainty principle, for dealing with signal reconstruction in communications theory. We are back to the raison d'être of the *ANHA* series!

John J. Benedetto
Series Editor
University of Maryland
College Park

Preface

The topic of this book is the design of sequences with good correlation properties. The ideas presented in the work evolved, in part, out of a radar project whose goal was to develop algorithms for processing linear frequency modulated (FM) chirp pulse reflections from multiple targets and dielectric materials. An especially convenient framework for this task is given by one of the time-frequency spaces, the Zak space. The Zak space provides a natural setting for studying chirps, as they are intrinsically time-frequency signals and have simple realizations as modulated algebraic lines on the Zak transform lattice. Because the replacement of time-domain analysis with two-dimensional Zak space analysis leads to the replacement of one-dimensional reflections with two-dimensional images, image processing techniques can be used to estimate target parameters, including dielectric material properties.

Complex target discrimination often requires the use of multi-beam imaging schemes. Multi-beam imaging can also be applied to aid in multi-material identification and in noise, multi-path and atmospheric interference rejection. To realize this advantage, the information contained in individual components of the echo must be unambiguously retrieved at the receiver. At the same time, to maintain an acceptable receiver signal-to-noise ratio, a sequence set must make maximum use of the available bandwidth. An accommodation of these two constraints necessitates identification of a sequence set that is sufficiently large and whose individual members interfere with each other as little as possible.

We address these signal design problems by Zak space methods. Zak space methods are used first to prove several well-known correlation properties of sampled linear FM chirp pulses. Among the theoretical results, we show that the Zak space representation of a discrete linear FM chirp is the matrix product of several copies of a permutation matrix and a diagonal matrix. This factorization decouples the permutation matrix from the modulating diagonal matrix. This result is the key to understanding the central role played by the Zak transform in our approach to sequence design. It permits the permutation group to take a central role in the design. Because certain classes of chirps are identical with certain communications sequences, obtained by a direct, time-domain approach, several results between the two intersect. However, the approach described in this work is broader in that it offers a novel way of analyzing chirps by both algebraic and geometric means and leads to the construction of much more general signal sets.

Along with mathematical developments illustrated by numerous examples, the text contains many tables. These tables compare the number and size of signal sets satisfying good correlation properties of specified periodicity attainable by the methods of communication theory with those attainable by Zak space methods. While the theoretical benefit of the Zak space approach is demonstrated, the efficacy of the approach largely depends on how, and to what advantage, the newly designed sequences can be used in applications. To address this question, several chapters describe aperiodic correlation properties and methods for limiting bandwidth and smoothing of analog envelopes. In the last chapter, a list of open problems is given and directions for further research are discussed.

We owe a good deal to Dr. Albanese of Brooks City-Base whose patience and keen insights provided the motivation for many of the results and, more importantly, the emphasis in the text. Through this collaboration, the mathematics of chirps and sequence design are never far removed from significant applications to radar image processing and materials identification. We also thank Dr. J. Oeschger of the Naval Surface Warfare Center who introduced us to interferometric synthetic aperture sonar image processing and requirements on sequence sets. The design of real-valued sequence sets in Chapter 15 is mainly due to this collaboration. Lastly, we are grateful to Dr. Rushanan and the late Dr. Fante, both of The MITRE Corporation, for discussions of multiple access communication systems and multi-beam radar applications.

Myong An
Andrzej K. Brodzik
Richard Tolimieri

April 2008

Contents

1

Introduction

In this text new methods for studying the correlation properties of polyphase sequences are developed. These methods are applied to the design of large collections of polyphase sequence pairs satisfying ideal correlation and polyphase sequence sets satisfying pairwise ideal correlation. Throughout the text a sequence is said to be polyphase if its components have equal absolute value.

Our approach is based on the finite Zak transform, and the point of departure in the investigation is the discrete linear frequency modulated (FM) chirp. Chirps are intrinsically time-frequency signals and have sparse and highly structured representations in time-frequency spaces. The correlation properties of quadratic chirps have been studied in [1]. We will prove several of these results using Zak space methods as motivation for the sequence design procedures developed in this text. Among the main theoretical results, we show that the finite Zak transform of a discrete chirp is the product of several copies of a permutation matrix and a diagonal matrix whose nonzero entries have absolute value one. This result is the key to understanding the central role played by the Zak transform in our approach to sequence design. This approach is based on designing sequences directly in the Zak space, with separate focus given to permutations and modulations. In the first part of the book, we define a new class of sequences, the permutation sequences, by the condition that their Zak space representations are permutation matrices. The permutation sequences automatically satisfy ideal autocorrelation. However, a pair of permutation sequences does not necessarily satisfy ideal cross correlation, requiring further analysis. The cornerstone of this analysis is the identification of a special class of permutations, the ∗-permutations. Each ∗-permutation serves as a root for a large class of permutation sequence pairs satisfying ideal correlation. As part of this investigation, we distinguish the class of discrete chirps, which up to modulation are ∗-permutation sequences. In all cases the collection of sequence pairs determined by a root can be algebraically computed. The characterization of permutation sequence pairs satisfying ideal correlation is the main result needed to construct sequence sets of modulated permutation sequences satisfying pairwise ideal correlation.

In the second part of the book, we bring back the diagonal matrix factor. A modulated permutation sequence is a sequence whose Zak transform is a permutation

M. An et al., *Ideal Sequence Design in Time-Frequency Space*,
DOI 10.1007/978-0-8176-4738-4_1,
© Birkhäuser Boston, a part of Springer Science+Business Media, LLC 2009

matrix or several copies of a permutation matrix multiplied by a diagonal matrix whose diagonal entries have absolute value one. Modulating a permutation sequence changes the shape of the sequence but not its cyclic correlation properties. A generalization of a permutation sequence is a sequence whose Zak transform consists of more than one permutation matrix. This sequence does not satisfy ideal autocorrelation. In this case modulation is required to impose ideal autocorrelation.

The main tool for studying the correlation properties of permutation sequences and modulated permutation sequences is the Zak space correlation formula. This formula describes the Zak space representation of the correlation of two sequences in terms of componentwise products of the columns of the Zak transform of the two sequences. Underlying the correlation formula is the relationship between sequence shifts and the Zak transform of sequence shifts. This formula, together with the Zak transform factorization result, guides the construction of new sequences.

Although we believe the Zak space sequence design approach presented here is original, we are aware of and indebted to previous efforts. The nature of the discrete chirp as a time-frequency signal was first recognized by Lerner [28]. The importance of the Zak transform in the study of the continuous chirp, as well as its special role in signal processing, was first pointed out by Janssen [25]. Janssen also derived the main properties of the Zak transform, including the Zak space correlation formula, and determined the Fourier transform and the Zak transform of the continuous chirp. There is also a wealth of literature in the use of other time-frequency transforms for studying the properties of discrete FM chirps [3, 35].

Although the main motivation for the approach in the text comes from sonar and radar systems, there is a common link to several works in communications sequence design. The link to finite field sequence design [11, 12, 15, 20, 21, 22, 23, 27, 31] is less clear, but the concept of a ∗-permutation introduced in this text has similarities with that of shift sequences [18]. The important text by S.W. Golomb and G. Gong [20] contains a treatment of shift sequences and their application to the design of sequences having good correlation properties.

The presentation of the material is self-contained, and background information is given when needed. In particular, as the constructions rely on matrix algebra, tensor products and permutation groups, a brief review of these topics is provided in Chapters 2 and 3. In Chapters 4 and 5 the main digital signal processing concepts, the finite Fourier transform and the correlation, are developed. Chapters 6 through 9 introduce the discrete chirp and the finite Zak transform, and state the main results on Zak space representations of chirps, including the Zak space correlation of chirps. Chapters 14 through 15 form the core of the book, as they develop the Zak space design framework of ideal sequences. Chapter 10 characterizes the set of ∗-permutations that are associated with ideal permutation sequences. Chapter 11 analyzes properties of permutation sequences based on ∗-permutations and identifies several families of ideal sequences. Chapter 12 investigates properties of modulation and derives the condition for arbitrary sequences to satisfy ideal correlation. In Chapter 13 we use results on permutation sequence pairs to develop design strategies for constructing collections of permutation sequences satisfying pairwise ideal correlations. Some of these strategies lead to explicit construction, but perhaps the most interesting result is numerical procedures for constructing large numbers of such collections. The numerics for several sizes of sequence collections are given in tables. Chapters 14 and 15 address several engineering issues pertinent to radar and sonar signal processing. Chapter 16 outlines several outstanding time-frequency sequence design problems.

2

Review of Algebra

In this chapter we introduce some of the notation and algebra used in this text. References include [24], [40]. Other results will be discussed as needed. Throughout this work $N > 1$ is an integer.

2.1 Ring of Integers

$\mathbb{Z}/N$ is the set

$$\{0, 1, \ \ldots, \ N - 1\}$$

under addition and multiplication modulo N. $\mathbb{Z}/N$ is a commutative ring called the *ring of integers modulo* N. 0 is the identity for addition and 1 is the identity for multiplication. For an arbitrary integer r,

$$r \bmod N$$

is the unique integer in $\mathbb{Z}/N$ equal to r modulo N.

We write

$$(r, N) = 1$$

to mean r is *relatively prime* to N. U_N is the set of integers in $\mathbb{Z}/N$ that are invertible under the multiplication in $\mathbb{Z}/N$. U_N is a group under multiplication modulo N called the *group of units* of $\mathbb{Z}/N$,

$$U_N = \{r \in \mathbb{Z}/N : (r, N) = 1\}.$$

In the next chapter we identify the group of units, U_N, of $\mathbb{Z}/N$ with a special subgroup of $N \times N$ permutation matrices, the group of unit permutation matrices. The unit permutation matrices play an essential part in the Zak space representation of discrete chirps in Chapter 9.

M. An et al., *Ideal Sequence Design in Time-Frequency Space*,
DOI 10.1007/978-0-8176-4738-4_2,
© Birkhäuser Boston, a part of Springer Science+Business Media, LLC 2009

Example 2.1

$$U_5 = \{1,\ 2,\ 3,\ 4\}.$$
$$U_8 = \{1,\ 3,\ 5,\ 7\}.$$
$$U_9 = \{1,\ 2,\ 4,\ 5,\ 7,\ 8\}.$$
$$U_{15} = \{1,\ 2,\ 4,\ 7,\ 8,\ 11,\ 13,\ 14\}.$$

In general the ratio

$$\frac{|U_N|}{N}$$

does not grow as N grows. This ratio affects the maximum order of sequence sets having good correlation properties of size N.

For a prime p, $\mathbb{Z}/p$ is a field and

$$U_p = \{1,\ \ldots,\ p-1\}.$$

U_p is a cyclic group under multiplication modulo p. This means that there exists $m \in U_p$ such that

$$U_p = \{m^j : 0 \le j < p-1\}.$$

m is called a *generator* (it is not unique) of U_p. In general the set of nonzero elements of a finite field is a cyclic group under field multiplication. If p is an odd prime and $r \ge 1$ is an integer, then

$$U_{p^r}$$

is a cyclic group, having $p^{r-1}(p-1)$ elements.

$$U_{p^r} = \{m^j : 0 \le j < p^{r-1}(p-1)\}.$$

We call m a generator of U_{p^r}.

Example 2.2

$$U_7 = \{1,\ 2,\ 3,\ 4,\ 5,\ 6\} = \left\{3^j : 0 \le j < 6\right\},$$

where 3^j is taken modulo 7, and $3^6 \equiv 1 \bmod 7$.

Example 2.3

$$U_9 = \{1,\ 2,\ 4,\ 5,\ 7,\ 8\} = \left\{2^j : 0 \le j < 6\right\},$$

where 2^j is taken modulo 9, and $2^6 \equiv 1 \bmod 9$.

For N not a prime power U_N is not a cyclic group, but by the Chinese remainder theorem U_N is the group direct product

$$U_N = U_{p_1^{r_1}} \times \cdots \times U_{p_t^{r_t}},$$

where $p_1^{r_1},\ \ldots,\ p_t^{r_t}$ are the distinct prime power factors of N.

Example 2.4 $U_{15} = U_3 \times U_5$.

Example 2.5 U_{p^2}, p an odd prime, consists of the p blocks of $p - 1$ integers

$$1, \ 2, \ \ldots, \ p - 1,$$

$$p + 1, \ p + 2, \ \ldots, \ 2p - 1,$$

$$\vdots$$

$$p(p - 1) + 1, \ p(p - 1) + 2, \ \ldots, \ p^2 - 1.$$

2.2 Vectors and Matrices

$\mathbb{C}$ is the complex field and $\mathbb{C}_1$ is the multiplicative group of complex numbers of absolute value 1. $\mathbb{C}^N$ is the complex vector space of all ordered N-tuples of complex numbers. A vector $\mathbf{x} \in \mathbb{C}^N$ is written

$$\mathbf{x} = [x_n]_{0 \leq n < N} = \begin{bmatrix} x_0 \\ x_1 \\ \vdots \\ x_{N-1} \end{bmatrix}.$$

The inner product of two vectors $\mathbf{x}$ and $\mathbf{y}$ in $\mathbb{C}^N$ is

$$\langle \mathbf{x}, \mathbf{y} \rangle = \sum_{n=0}^{N-1} x_n y_n^*,$$

where $*$ is complex conjugation. The norm of $\mathbf{x}$ is

$$\|\mathbf{x}\| = \sqrt{\langle \mathbf{x}, \mathbf{x} \rangle}.$$

The set of vectors

$$\{\mathbf{e}_n : 0 \leq n < N\},$$

where $\mathbf{e}_n$ is the vector in $\mathbb{C}^N$ having 1 in the n-th component and 0 in all other components is an orthonormal basis of $\mathbb{C}^N$

$$\langle \mathbf{e}_r, \mathbf{e}_s \rangle = \begin{cases} 1, \ r = s \\ 0, \ r \neq s, \end{cases} \quad 0 \leq r, s < N.$$

When it is not clear from context we write $\mathbf{e}_n^N$ or write $\mathbf{e}_n \in \mathbb{C}^N$ to indicate the appropriate space. A vector $\mathbf{x} \in \mathbb{C}^N$ can be uniquely written as

$$\mathbf{x} = \sum_{n=0}^{N-1} x_n \mathbf{e}_n, \quad x_n \in \mathbb{C}.$$

1 is the vector in $\mathbb{C}^N$ all of whose components are equal to 1 and **0** is the vector in $\mathbb{C}^N$ all of whose components are equal to 0. When it is not clear from context, we write $\mathbf{1}^N$ and $\mathbf{0}^N$ to indicate the appropriate space. The vector **1** can be written

$$\mathbf{1} = \sum_{n=0}^{N-1} \mathbf{e}_n.$$

Componentwise multiplication of vectors plays an important role in this text. The notation

$$\mathbf{xy}$$

denotes the vector in $\mathbb{C}^N$ formed by componentwise multiplication,

$$\mathbf{xy} = [x_n y_n]_{0 \le n < N}.$$

In particular,

$$\mathbf{e}_r \mathbf{e}_s = \begin{cases} \mathbf{e}_r, & r = s \\ \mathbf{0}, & r \ne s, \end{cases} \qquad 0 \le r, s < N,$$

and

$$\mathbf{e}_r \left(\sum_{n=0}^{N-1} x_n \mathbf{e}_n \right) = \mathbf{e}_r \mathbf{x} = x_r \mathbf{e}_r, \qquad 0 \le r < N.$$

L and K are positive integers. A complex $L \times K$ matrix X is written

$$X = [x_{l,k}]_{0 \le l < L, 0 \le k < K}.$$

We usually write a complex $L \times K$ matrix in terms of its column vectors,

$$[\mathbf{x}_0 \cdots \mathbf{x}_{K-1}], \qquad \mathbf{x}_k \in \mathbb{C}^L$$

or

$$[X_0 \cdots X_{K-1}], \qquad X_k \in \mathbb{C}^L.$$

For this reason we set

$$E_n = \mathbf{e}_n, \qquad 0 \le n < N.$$

The *trace* of an $N \times N$ matrix X is

$$\operatorname{Tr} X = \sum_{n=0}^{N-1} x_{n,n}.$$

The trace is defined only for square matrices.

Special matrices will be used throughout this text. We introduce them now, but study them more completely in subsequent chapters. I_N is the $N \times N$ identity matrix. The $N \times N$ *shift* matrix S defined by

$$Sx = \begin{bmatrix} x_{N-1} \\ x_0 \\ \vdots \\ x_{N-2} \end{bmatrix}$$

and the $N \times N$ *time reversal* matrix R defined by

$$Rx = \begin{bmatrix} x_0 \\ x_{N-1} \\ \vdots \\ x_1 \end{bmatrix}$$

are examples of permutation matrices. Permutation matrices will be studied in detail in Chapter 3. When dimension must be distinguished we write S_N for S and R_N for R. Several elementary results about S and R will be described at this time.

S has order N,

$$S^N = I_N,$$

and N is the smallest positive integer with this property. As a result S is invertible and

$$S^{-1} = S^{N-1}.$$

The componentwise product of any two distinct columns of S^n, $0 \le n < N$, is $\mathbf{0}$ and

$$\mathrm{Tr} \ S^n = 0, \quad 0 < n < N.$$

R has order 2,

$$R^2 = I_N.$$

The componentwise product of any two distinct columns of R is $\mathbf{0}$ and

$$Tr \ R = 1.$$

R and S are related by

$$RSR = S^{-1}.$$

The $N \times N$ Fourier transform matrix F is defined by

$$F = [w^{mn}]_{0 \le m,n < N}, \quad w = e^{2\pi i \frac{1}{N}}.$$

When dimension must be distinguished we write $F(N)$ for F. The $N \times N$ Fourier transform matrix will be studied in detail in Chapter 4. This will include the important relationships between S, R and F.

For $\mathbf{x} \in \mathbb{C}^N$, $D(\mathbf{x})$ is the $N \times N$ diagonal matrix defined by

$$D(\mathbf{x}) = \begin{bmatrix} x_0 & & & \\ & x_1 & & 0 \\ & & \ddots & \\ 0 & & & x_{N-1} \end{bmatrix}.$$

Componentwise multiplication and diagonal matrix multiplication can be inter-changed,

$$\mathbf{xy} = D(\mathbf{x})\mathbf{y}, \quad \mathbf{x}, \mathbf{y} \in \mathbb{C}^N.$$

An important example is

$$D = D\left([w^n]_{0 \le n < N}\right), \quad w = e^{2\pi i \frac{1}{N}}.$$

When dimension must be distinguished we write D_N for D.

D has order N,

$$D^N = I_N,$$

and N is the smallest positive integer having this property.

2.3 Tensor Products

Throughout this section $N = LK$, where $L > 1$, $K > 1$ and $M > 1$ are integers.

Tensor products of vectors and matrices can be found in many branches of physics and mathematics. They have been used in digital signal processing to model many results and algorithms in terms of matrix factorizations. Tensor products and stride permutations, introduced in the next chapter, form the basic building blocks of the tensor product algebra which can be used to describe complex computations and can serve as an interactive programming tool. A complete treatment of the tensor product algebra along with the tensor product identities and its application to algorithms for the finite Fourier transform and convolution is contained in [40] and [41].

In this work the tensor product algebra is more than an implementation tool. It is the algebra underlying derivations and results, replacing the summations and multiple indices usually found in digital signal processing texts.

The *tensor product* of vectors $\mathbf{x} \in \mathbb{C}^L$ and $\mathbf{y} \in \mathbb{C}^K$ is the vector $\mathbf{x} \otimes \mathbf{y} \in \mathbb{C}^N$ defined by

$$\mathbf{x} \otimes \mathbf{y} = \begin{bmatrix} x_0 \mathbf{y} \\ \vdots \\ x_{L-1} \mathbf{y} \end{bmatrix}.$$

We can view $\mathbf{x} \otimes \mathbf{y}$ as consisting of L contiguous vector segments each in $\mathbb{C}^K$,

$$x_l \mathbf{y}, \quad 0 \le l < L.$$

Example 2.6

$$\begin{bmatrix} x_0 \\ x_1 \\ x_2 \end{bmatrix} \otimes \begin{bmatrix} y_0 \\ y_1 \end{bmatrix} = \begin{bmatrix} x_0 y_0 \\ x_0 y_1 \\ x_1 y_0 \\ x_1 y_1 \\ x_2 y_0 \\ x_2 y_1 \end{bmatrix}.$$

The collection of tensor products

$$\mathbf{x} \otimes \mathbf{y}, \qquad \mathbf{x} \in \mathbb{C}^L, \mathbf{y} \in \mathbb{C}^K,$$

spans $\mathbb{C}^N$, but is not a basis.

The tensor product is associative and distributive,

$$(\mathbf{x} \otimes \mathbf{y}) \otimes \mathbf{z} = \mathbf{x} \otimes (\mathbf{y} \otimes \mathbf{z}),$$

$$(\mathbf{x} + \mathbf{y}) \otimes \mathbf{z} = \mathbf{x} \otimes \mathbf{z} + \mathbf{y} \otimes \mathbf{z}$$

and

$$\mathbf{x} \otimes (\mathbf{y} + \mathbf{z}) = \mathbf{x} \otimes \mathbf{y} + \mathbf{x} \otimes \mathbf{z}, \qquad \mathbf{x} \in \mathbb{C}^L, \mathbf{y} \in \mathbb{C}^K, \mathbf{z} \in \mathbb{C}^M,$$

but is not commutative. We also have

$$(\alpha \mathbf{x}) \otimes \mathbf{y} = \alpha(\mathbf{x} \otimes \mathbf{y}) = \mathbf{x} \otimes (\alpha \mathbf{y}), \qquad \alpha \in \mathbb{C}.$$

These results can be derived by following the definition of the tensor product.

Example 2.7 $e_r^L \otimes e_s^K = e_{s+rK}^N$.

$T = RS$, where $R > 1$, $S > 1$ are integers. The *tensor product* of an $L \times R$ matrix X and a $K \times S$ matrix Y is the $N \times T$ matrix $X \otimes Y$ defined by

$$X \otimes Y = [x_{l,r}Y]_{0 \le l < L, 0 \le r < R} .$$

We can view $X \otimes Y$ as consisting of LR matrix blocks

$$x_{l,r}Y, \qquad 0 \le l < L, 0 \le r < R,$$

each a $K \times S$ matrix.

Example 2.8 $I_2 \otimes Y = Y \oplus Y = \begin{bmatrix} Y & 0 \\ 0 & Y \end{bmatrix}$, where I_2 is the 2×2 identity matrix and $\oplus$ is the matrix direct sum.

In general $I_L \otimes Y$ is the block diagonal matrix having L copies of Y along the diagonal and 0 elsewhere.

Example 2.9

$$\begin{bmatrix} a & b \\ c & d \end{bmatrix} \otimes I_2 = \begin{bmatrix} a & 0 & b & 0 \\ 0 & a & 0 & b \\ c & 0 & d & 0 \\ 0 & c & 0 & d \end{bmatrix}.$$

Write the $L \times R$ matrix X and the $K \times S$ matrix Y in terms of their column vectors,

$$X = [\mathbf{x}_0 \ \cdots \ \mathbf{x}_{R-1}] \ \text{and} \ Y = [\mathbf{y}_0 \ \cdots \ \mathbf{y}_{s-1}].$$

Then

$$\mathbf{x}_r \otimes Y = [\mathbf{x}_r \otimes \mathbf{y}_0 \ \cdots \ \mathbf{x}_r \otimes \mathbf{y}_{s-1}]$$

and

$$X \otimes Y = [\mathbf{x}_0 \otimes Y \ \cdots \ \mathbf{x}_{r-1} \otimes Y].$$

These formulas can be used to show that the tensor product of matrices is associative and distributive, but not commutative. We also have

$$(X \otimes Y)(\mathbf{x} \otimes \mathbf{y}) = X\mathbf{x} \otimes Y\mathbf{y}$$

and

$$(X \otimes Y)(Z \otimes W) = XZ \otimes YW$$

for appropriate size matrices X, Y, Z, W and vectors $\mathbf{x}$ and $\mathbf{y}$. As a consequence, if X and Y are invertible,

$$(X \otimes Y)^{-1} = X^{-1} \otimes Y^{-1}.$$

Example 2.10 $S_{2L}^2 = S_L \otimes I_2$.

In general

$$S_N^K = S_L \otimes I_K.$$

3

Permutations and Permutation Matrices

Permutations play several important roles in this text. They are essential for the Zak space description of discrete chirps derived in Chapter 9 and in the definition of permutation sequences in Chapter 11. The set of permutation matrices is a group under matrix multiplication. The group property is used, beginning in Chapter 11, to give conditions for ideal correlation.

Special classes of permutation matrices will be defined. The unit permutation matrices are used to describe Zak space representations of discrete chirps. The stride permutations complete the tensor product algebra and give rise to the tensor product identities. Many constructions and implementations can be formulated in terms of the tensor product algebra.

In Chapter 10 the important set of $*$-permutations is defined. Subsequent chapters construct ideal correlation signal pairs in terms of $*$-permutations.

Throughout this chapter $L > 1$ is an integer. $Perm(L)$ is the collection of all one-to-one mappings of $\mathbb{Z}/L$ onto itself. We write Λ for $Perm(L)$. When the dependence on L must be distinguished we write $\Lambda(L)$ for Λ.

Λ is a noncommutative group, called the *permutation group*, under the composition of the mappings. Its order is $L!$ For $\phi_1, \phi_2 \in \Lambda$, the product $\phi_1\phi_2 \in \Lambda$ is defined by

$$(\phi_1\phi_2)(r) = \phi_1(\phi_2(r)), \quad r \in \mathbb{Z}/L.$$

The identity permutation is denoted by ϕ_0,

$$\phi_0(r) = r, \quad r \in \mathbb{Z}/L.$$

ϕ^{-1} is the inverse of $\phi \in \Lambda$.

A permutation $\phi \in \Lambda$ is written

$$\phi = (\phi(0) \quad \phi(1) \quad \cdots \quad \phi(L-1)).$$

For $\mathbf{n} \in (\mathbb{Z}/L)^L$ with pairwise distinct components define $\phi_\mathbf{n} \in \Lambda$ by

$$\phi_\mathbf{n} = (n_0 \quad n_1 \quad \cdots \quad n_{L-1}).$$

M. An et al., *Ideal Sequence Design in Time-Frequency Space*,
DOI 10.1007/978-0-8176-4738-4_3,
© Birkhäuser Boston, a part of Springer Science+Business Media, LLC 2009

Example 3.1 For $L = 5$ and $\phi_1 = (0 \ 2 \ 4 \ 1 \ 3)$, $\phi_2 = (0 \ 3 \ 1 \ 4 \ 2)$,

$$\phi_1 \phi_2 = \phi_0 = \phi_2 \phi_1.$$

Example 3.2 For $L = 5$ and $\phi_1 = (0 \ 4 \ 3 \ 1 \ 2)$, $\phi_2 = (0 \ 3 \ 1 \ 4 \ 2)$,

$$\phi_1 \phi_2 = (0 \ 1 \ 4 \ 2 \ 3) \neq \phi_2 \phi_1 = (0 \ 2 \ 4 \ 3 \ 1)$$

and

$$\phi_1^{-1} = (0 \ 3 \ 4 \ 2 \ 1).$$

We will add and subtract permutations. More generally, for mappings ϕ_1 and ϕ_2 of $\mathbb{Z}/L$ *into* itself, define the mapping $\phi_1 + \phi_2$ of $\mathbb{Z}/L$ into itself by

$$(\phi_1 + \phi_2)(r) = \phi_1(r) + \phi_2(r), \quad r \in \mathbb{Z}/L.$$

The sum on the right-hand side is taken modulo L. Even when ϕ_1 and ϕ_2 are permutations, the sum $\phi_1 + \phi_2$ is not necessarily a permutation. The difference of mappings is defined in a similar way. It is important to note that for mappings ϕ_1, ϕ_2 and ϕ of $\mathbb{Z}/L$ into itself,

$$(\phi_1 + \phi_2)\phi = \phi_1 \phi + \phi_2 \phi,$$

but

$$\phi(\phi_1 + \phi_2) \neq \phi\phi_1 + \phi\phi_2.$$

Example 3.3 For $L = 7$ and

$$\phi_1 = (0 \ 2 \ 4 \ 6 \ 1 \ 3 \ 5), \quad \phi_2 = (0 \ 3 \ 6 \ 2 \ 5 \ 1 \ 4),$$

$$\phi_1 + \phi_2 = (0 \ 5 \ 3 \ 1 \ 6 \ 4 \ 2)$$

and

$$\phi_1 - \phi_2 = (0 \ 6 \ 5 \ 4 \ 3 \ 2 \ 1).$$

Both $\phi_1 + \phi_2$ and $\phi_1 - \phi_2$ are permutations.

Example 3.4 For $L = 7$ and

$$\phi_1 = (0 \ 2 \ 3 \ 6 \ 1 \ 4 \ 5), \quad \phi_2 = (0 \ 3 \ 6 \ 2 \ 5 \ 1 \ 4),$$

$$\phi_1 + \phi_2 = (0 \ 5 \ 2 \ 1 \ 6 \ 5 \ 2)$$

$$\phi_1 - \phi_2 = (0 \ 6 \ 4 \ 4 \ 3 \ 3 \ 1).$$

Neither $\phi_1 + \phi_2$ nor $\phi_1 - \phi_2$ are permutations.

X is an $L \times L$ matrix which we write as

$$X = [X_0 \ \cdots \ X_{L-1}].$$

For a mapping ϕ of $\mathbb{Z}/L$ into itself, X_ϕ is the $L \times L$ matrix defined by

$$X_\phi = \begin{bmatrix} X_{\phi(0)} & \cdots & X_{\phi(L-1)} \end{bmatrix}.$$

If ϕ is a permutation, then X_ϕ is formed by permuting the columns of X. For a mapping ϕ of $\mathbb{Z}/L$ into itself, E_ϕ is the $L \times L$ matrix defined by

$$E_\phi = \begin{bmatrix} E_{\phi(0)} & \cdots & E_{\phi(L-1)} \end{bmatrix}.$$

If ϕ is a permutation, then E_ϕ is a *permutation matrix*. Since

$$E_\phi^{-1} = F_{\phi^{-1}}$$

$$E_{\phi_1 \phi_2} = E_{\phi_1} E_{\phi_2} \quad \phi, \phi_1, \phi_2 \in \Lambda,$$

the set of all permutation matrices

$$\{E_\phi : \phi \in \Lambda\}$$

is a group under matrix multiplication.

Several elementary properties of permutation matrices follow. For $\phi \in \Lambda$

$$F_\phi \mathbf{x} = \begin{bmatrix} x_{\phi^{-1}(l)} \end{bmatrix}_{0 \le l < L}, \quad \mathbf{x} \in \mathbb{C}^L,$$

$$E_\phi^T = E_\phi^{-1}, \quad T \text{ the transpose}.$$

For an $L \times L$ matrix X

$$X E_\phi = X_\phi,$$

and for $\mathbf{x} \in \mathbb{C}^L$

$$E_\phi D(\mathbf{x}) E_\phi^{-1} = D(E_\phi \mathbf{x}).$$

In the following chapters we repeatedly use the fact that the componentwise product of any column vector of a permutation matrix with itself is the column vector and the fact that the componentwise product of two distinct columns of a permutation matrix is the zero vector.

Example 3.5 For $L = 5$ and $\phi = (1 \ 2 \ 3 \ 4 \ 0)$,

$$E_\phi = \begin{bmatrix} 0 & 0 & 0 & 0 & 1 \\ 1 & 0 & 0 & 0 & 0 \\ 0 & 1 & 0 & 0 & 0 \\ 0 & 0 & 1 & 0 & 0 \\ 0 & 0 & 0 & 1 & 0 \end{bmatrix} = S_5,$$

and

$$E_\phi \mathbf{x} = \begin{bmatrix} x_4 \\ x_0 \\ x_1 \\ x_2 \\ x_3 \end{bmatrix},$$

where $\phi^{-1} = (4 \ 0 \ 1 \ 2 \ 3)$.

Example 3.6 For $L = 5$ and $\phi = (0\ \ 2\ \ 4\ \ 1\ \ 3)$,

$$E_\phi = \begin{bmatrix} 1 & 0 & 0 & 0 & 0 \\ 0 & 0 & 0 & 1 & 0 \\ 0 & 1 & 0 & 0 & 0 \\ 0 & 0 & 0 & 0 & 1 \\ 0 & 0 & 1 & 0 & 0 \end{bmatrix}$$

and

$$X E_\phi = [X_0\ \ X_2\ \ X_4\ \ X_1\ \ X_3].$$

Example 3.7 For $L = 6$ and $\phi = (0\ \ 2\ \ 4\ \ 1\ \ 3\ \ 5)$,

$$E_\phi = \begin{bmatrix} 1 & 0 & 0 & 0 & 0 & 0 & 0 \\ 0 & 0 & 0 & 0 & 1 & 0 & 0 \\ 0 & 1 & 0 & 0 & 0 & 0 & 0 \\ 0 & 0 & 0 & 0 & 0 & 1 & 0 \\ 0 & 0 & 1 & 0 & 0 & 0 & 0 \\ 0 & 0 & 0 & 0 & 0 & 0 & 1 \\ 0 & 0 & 0 & 1 & 0 & 0 & 0 \end{bmatrix}.$$

In the following sections we will identify certain subgroups of permutations and permutation matrices that will, in following chapters, play a major role in the Zak space description of discrete chirps and in the design of sequence sets having good correlation properties.

3.1 Shift Permutations

The *shift permutation* $\sigma \in \Lambda$ is defined by

$$\sigma = (1\ \ 2\ \ \cdots\ \ L-1\ \ 0).$$

We can also write

$$\sigma(n) = n + 1, \qquad n \in \mathbb{Z}/L.$$

Example 3.8 For $L = 5$,

$$\sigma = (1\ 2\ 3\ 4\ 0),\quad \sigma^2 = (2\ 3\ 4\ 0\ 1),\quad \sigma^3 = (3\ 4\ 0\ 1\ 2),$$

$$\sigma^4 = (4\ 0\ 1\ 2\ 3),$$

and

$$\sigma^5 = (0\ 1\ 2\ 3\ 4) = \phi_0.$$

In general,

$$\sigma^L = \phi_0,$$

and L is the smallest positive power of σ with this property.

Shift permutations are related to shift matrices by

$$E_\sigma = S$$

and

$$E_{\sigma^r} = S^r, \qquad 0 \le r < L.$$

When the dependence of σ on L must be distinguished, we write σ_L for σ.

If $\phi \in \Lambda$, then

$$\phi\sigma = (\phi(1) \quad \cdots \quad \phi(L-1) \quad \phi(0))$$

and

$$\sigma\phi = (\phi(0) + 1 \quad \cdots \quad \phi(L-1) + 1).$$

Example 3.9 For $L = 5$ and $\phi = (0\ 2\ 4\ 1\ 3)$,

$$\phi\sigma = (2\ 4\ 1\ 3\ 0)$$

and

$$\sigma\phi = (1\ 3\ 0\ 2\ 4).$$

In particular, $\phi\sigma \ne \sigma\phi$.

For an $L \times L$ matrix X,

$$XE_\sigma = X_\sigma = XS.$$

XS is formed by shifting the columns of X by σ. On the other hand, SX shifts the rows of X by σ.

3.2 Unit Permutations

In Chapter 2 we introduced the multiplicative group U_L of units in the ring $\mathbb{Z}/L$ of integers modulo L. Writing

$$(u, L) = 1, \qquad u \in \mathbb{Z},$$

to mean that u and L are relatively prime, we have

$$U_L = \{u \in \mathbb{Z}/L : (u, L) = 1\}.$$

1 is the identity in U_L and u^{-1} is the inverse of $u \in U_L$, and uv is the product of $u, v \in U_L$.

In this section we identify U_L with a commutative subgroup of Λ. The permutations identified with the elements of U_L appear, usually implicitly, in many previous works on polyphase sequence design [31].

For $u \in \mathbb{Z}/L$, μ_u is the mapping of $\mathbb{Z}/L$ into itself defined by

$$\mu_u(r) = ur, \quad r \in \mathbb{Z}/L.$$

$\mu_1 = \phi_0$. μ_u is a permutation if and only if $u \in U_L$. In this case we call μ_u a *unit permutation*. For unit permutations

$$\mu_u^{-1} = \mu_{u^{-1}}$$

and

$$\mu_u\mu_v = \mu_{uv} = \mu_v\mu_u.$$

Set Λ_0 equal to the set of all unit permutations,

$$\Lambda_0 = \{\mu_u : u \in U_L\}.$$

By the relations above, Λ_0 is a commutative subgroup of Λ. The arithmetic of multiplying and inverting unit permutations is identical to that in U_L. When the dependence of Λ_0 on L must be distinguished, we write $\Lambda_0(L)$ for Λ_0.

Example 3.10 $\Lambda_0(4) = \{\mu_1, \mu_3\}$.

$$\mu_3^2 = \mu_1.$$

Example 3.11 $\Lambda_0(5) = \{\mu_1, \mu_2, \mu_3, \mu_4\}$.

$$\mu_2^2 = \mu_4, \quad \mu_2\mu_3 = \mu_1, \quad \mu_2\mu_4 = \mu_3, \quad \mu_3^2 = \mu_4, \quad \mu_3\mu_4 = \mu_2, \quad \mu_4^2 = \mu_1.$$

Example 3.12 For $L = 12$, $\mu_2 = (0\ 2\ 4\ 6\ 8\ 10\ 0\ 2\ 4\ 6\ 8\ 10)$ is not a permutation while $\mu_5 = (0\ 5\ 10\ 3\ 8\ 1\ 6\ 11\ 4\ 9\ 2\ 7)$ is a permutation.

Example 3.13 For L odd, $\mu_2 \in \Lambda_0(L)$.

Example 3.14 For $L = p$, an odd prime,

$$\Lambda_0(p) = \{\mu_u : 1 \leq u < p\}.$$

As discussed in Chapter 2 $\Lambda_0(p)$ and $\Lambda_0(p^r)$, p an odd prime, are cyclic groups of order $p-1$ and $p^{r-1}(p-1)$. The group of unit permutations for composite L can be found using the Chinese remainder theorem which identifies the group U_{pq}, p,q distinct primes, for example, with the direct product of groups $U_p \times U_q$. Details can be found in [24, 40]. Using the Chinese remainder theorem identification the order of $\Lambda_0(pq)$ is $(p-1)(q-1)$. This identification plays no role in the theoretic aspects of this text, but is essential for constructing examples.

Example 3.15 The order of $\Lambda_0(15)$ is 8.

Mappings from $\mathbb{Z}/L$ into itself can be added and subtracted. In particular,

$$\mu_u + \mu_v = \mu_{u+v}, \qquad u, \ v \in \mathbb{Z}/L.$$

If u, v and $u - v$ are in U_L, then $\mu_u - \mu_v$ is a unit permutation,

$$\mu_u - \mu_v = \mu_{u-v} = \mu_u \left(\mu_1 - \mu_u^{-1} \mu_v \right).$$

Distribution on the right is allowed because unit permutations commute. This observation plays a critical role in subsequent chapters.

By an *affine permutation* we mean a product of the form

$$\sigma^l \mu_u, \quad l \in \mathbb{Z}/L, \quad u \in U_L.$$

The action of $\sigma^l \mu_u$ on $\mathbb{Z}/L$ is

$$\sigma^l \mu_u(r) = ur + l, \quad r \in \mathbb{Z}/L.$$

Affine permutations are usually required to describe the Zak space representation of discrete chirps.

The collection of affine permutations

$$A = \{ \sigma^l \mu_u : l \in \mathbb{Z}/L, \quad u \in U_L \}$$

is a subgroup of Λ, called the *affine group*. To see this we must show that the product of two affine permutations is an affine permutation. This follows from

$$(\sigma^l \mu_u)(\sigma^m \mu_v) = \sigma^n \mu_w, \qquad u, v \in U_L, \quad l, \ m \in \mathbb{Z}/L,$$

where $w = uv$ and $n = l + um$.

Λ_0 and the cyclic group generated by σ are commutative subgroups of A. However, because

$$\mu_u \sigma^m \mu_u^{-1} = \sigma^{um}, \qquad u \in U_L, \quad m \in \mathbb{Z}/L,$$

A is *not* a commutative group.

Example 3.16 $L = 5$.

	μ_2	μ_2^{-1}	σ	$\mu_2 \sigma \mu_2^{-1} = \sigma^2$
0	0	0	1	2
1	2	3	2	3
2	4	1	3	4
3	1	4	4	0
4	3	2	0	1

The permutation matrices corresponding to the unit permutations

$$E_{\mu_u}, \quad u \in U_L,$$

are called *unit permutation matrices*. Unit permutation matrices, viewed as an image in $L \times L$ space, are sometimes called *algebraic lines* through the *origin* of slope u in $L \times L$ space. The affine group permutation matrices

$$E_{\sigma^n \mu_u}, \quad n \in \mathbb{Z}/L, \quad u \in U_L,$$

describe *shifted* algebraic lines. The importance of these remarks will be made clear when we consider the echoes of linear FM discrete chirps.

Example 3.17 For $L = 7$,

$$E_{\mu_2} = \begin{bmatrix} 1 & 0 & 0 & 0 & 0 & 0 & 0 \\ 0 & 0 & 0 & 0 & 1 & 0 & 0 \\ 0 & 1 & 0 & 0 & 0 & 0 & 0 \\ 0 & 0 & 0 & 0 & 0 & 1 & 0 \\ 0 & 0 & 1 & 0 & 0 & 0 & 0 \\ 0 & 0 & 0 & 0 & 0 & 0 & 1 \\ 0 & 0 & 0 & 1 & 0 & 0 & 0 \end{bmatrix}, \quad E_{\sigma \mu_2} = \begin{bmatrix} 0 & 0 & 0 & 1 & 0 & 0 & 0 \\ 1 & 0 & 0 & 0 & 0 & 0 & 0 \\ 0 & 0 & 0 & 0 & 1 & 0 & 0 \\ 0 & 1 & 0 & 0 & 0 & 0 & 0 \\ 0 & 0 & 0 & 0 & 0 & 1 & 0 \\ 0 & 0 & 1 & 0 & 0 & 0 & 0 \\ 0 & 0 & 0 & 0 & 0 & 0 & 1 \end{bmatrix}.$$

Example 3.18 For $L = 25$, E_{μ_3} and $E_{\sigma^5 \mu_3}$ are viewed as images in Figure 3.1.

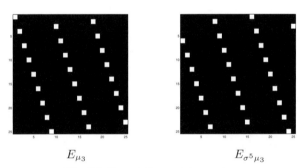

$$E_{\mu_3} \qquad\qquad\qquad E_{\sigma^5 \mu_3}$$

Fig. 3.1. Permutation matrices as images

3.3 Stride Permutations

$N = LK$, where $L > 1$ and $K > 1$ are integers.

Stride permutations occur in many application areas, including digital signal processing, error-correcting coding and finite field signal design. One reason, perhaps the major reason, is the relationship between decimated data and periodic data under the Fourier transform [41].

Stride permutations are, along with tensor products of vectors and matrices, basic building blocks of the tensor product algebra and are essential for describing the tensor product identities [40]. From this point of view stride permutations commute tensor products.

Suppose $\mathbf{x} \in \mathbb{C}^N$. Define the vectors $\mathbf{x}_k \in \mathbb{C}^L, 0 \leq k < K$, by

$$\mathbf{x}_k = [x_{k+lK}]_{0 \leq l < L}, \qquad 0 \leq k < K.$$

The vectors $\mathbf{x}_k, 0 \leq k < K$, are called the K-decimated components of $\mathbf{x}$. The N-point stride K permutation matrix P is defined by

$$P\mathbf{x} = \begin{bmatrix} \mathbf{x}_0 \\ \vdots \\ \mathbf{x}_{K-1} \end{bmatrix}.$$

When the dependence of P on K must be distinguished, we write P as $P(N, K)$.

Identify $\mathbb{C}^L \times \mathbb{C}^K$ with the space of $L \times K$ complex matrices. Define the $L \times K$ matrix $M\mathbf{x}$ by

$$M\mathbf{x} = [\mathbf{x}_0 \quad \cdots \quad \mathbf{x}_{K-1}].$$

M is called the $L \times K$ interleaving transform. When the dependence of M on L must be distinguished, we write M as M_L.

Example 3.19 For $L = 2$ and $K = 3$,

$$\mathbf{x}_0 = \begin{bmatrix} x_0 \\ x_3 \end{bmatrix}, \qquad \mathbf{x}_1 = \begin{bmatrix} x_1 \\ x_4 \end{bmatrix}, \qquad \mathbf{x}_2 = \begin{bmatrix} x_2 \\ x_5 \end{bmatrix}.$$

$$M\mathbf{x} = \begin{bmatrix} x_0 \ x_1 \ x_2 \\ x_3 \ x_4 \ x_5 \end{bmatrix}.$$

Example 3.20

$$P(6, 3) = \begin{bmatrix} 1 & 0 & 0 & 0 & 0 & 0 \\ 0 & 0 & 0 & 1 & 0 & 0 \\ 0 & 1 & 0 & 0 & 0 & 0 \\ 0 & 0 & 0 & 0 & 1 & 0 \\ 0 & 0 & 1 & 0 & 0 & 0 \\ 0 & 0 & 0 & 0 & 0 & 1 \end{bmatrix} = \mu_2.$$

The interleaving transform is the two-dimensional realization of the stride permutation. Both are used to describe the data reindexing in Cooley–Tukey fast Fourier transform algorithms [40, 41] and as a tool in error-correcting coding for constructing product codes and for embedding specialized short codes in long codes [5]. In relatively recent works [20, 21] the interleaving transform determines a powerful framework for unifying and generalizing large classes of finite field sequences having good correlation properties.

The interleaving operator M is the first step in the construction of the Zak transform defined in Chapter 7. Zak space is the main image representation space in this text. The relationship between the shift matrix, the interleaving transform and the Zak transform underlies all results and applications of the Zak transform from sequence design for good correlation to echo analysis.

We begin by describing the role of stride permutations in the tensor product algebra. The main result is that stride permutations commute tensor products. Details as to derivation and application to the modification of data rearrangements in a computation can be found in many places, including [40].

Suppose $\mathbf{x} \in \mathbb{C}^L$, $\mathbf{y} \in \mathbb{C}^K$. X and Y are $L \times L$ and $K \times K$ matrices. If $P = P(N, K)$, then

$$P(\mathbf{x} \otimes \mathbf{y}) = \mathbf{y} \otimes \mathbf{x}$$

and

$$P(X \otimes Y)P^{-1} = Y \otimes X.$$

As mentioned, tensor products have natural block structure. Stride permutations interchange block structure.

Because

$$P(N, L)(\mathbf{y} \otimes \mathbf{x}) = \mathbf{x} \otimes \mathbf{y}$$

we have

$$P(N, K)P(N, L)(\mathbf{y} \otimes \mathbf{x}) = \mathbf{y} \otimes \mathbf{x}.$$

The vector tensor products $\mathbf{x} \otimes \mathbf{y}$, $\mathbf{x} \in \mathbb{C}^L$, $\mathbf{y} \in \mathbb{C}^K$, span $\mathbb{C}^N$, implying

$$P^{-1} = P(N, L).$$

We begin with some examples describing the relationship between the interleaving operation and the shift matrices.

Example 3.21 For $L = 2$ and $K = 3$, set $M = M_2$ and $S = S_2$. For $\mathbf{x} \in \mathbb{C}^6$

$$M\mathbf{x} = \begin{bmatrix} x_0 \ x_1 \ x_2 \\ x_3 \ x_4 \ x_5 \end{bmatrix} = \begin{bmatrix} \mathbf{x_0} \ \mathbf{x_1} \ \mathbf{x_2} \end{bmatrix},$$

$$S_6\mathbf{x} = \begin{bmatrix} x_5 \\ x_0 \\ x_1 \\ x_2 \\ x_3 \\ x_4 \end{bmatrix}.$$

Direct computation shows

$$M(S_6\mathbf{x}) = \begin{bmatrix} x_5 \ x_0 \ x_1 \\ x_2 \ x_3 \ x_4 \end{bmatrix} = \begin{bmatrix} S\mathbf{x_2} \ \mathbf{x_0} \ \mathbf{x_1} \end{bmatrix}.$$

Example 3.22 For $L = 3$ and $K = 2$, set $M = M_3$ and $S = S_3$. For $\mathbf{x} \in \mathbb{C}^6$

$$M\mathbf{x} = \begin{bmatrix} x_0 \ x_1 \\ x_2 \ x_3 \\ x_4 \ x_5 \end{bmatrix} = \begin{bmatrix} \mathbf{x}_0 \ \mathbf{x}_1 \end{bmatrix},$$

and

$$S_6\mathbf{x} = \begin{bmatrix} x_5 \\ x_0 \\ x_1 \\ x_2 \\ x_3 \\ x_4 \end{bmatrix}.$$

Direct computation shows

$$M\left(S_6\mathbf{x}\right) = \begin{bmatrix} x_5 \ x_0 \\ x_1 \ x_2 \\ x_3 \ x_4 \end{bmatrix} = \begin{bmatrix} S\mathbf{x}_1 \ \mathbf{x}_0 \end{bmatrix}.$$

Suppose $M = M_L$ and $S = S_L$. The general formula is

$$M(S_N\mathbf{x}) = \begin{bmatrix} S\mathbf{x}_{K-1} \ \mathbf{x}_0 \ \cdots \ \mathbf{x}_{K-2} \end{bmatrix}, \qquad \mathbf{x} \in \mathbb{C}^N.$$

The interleaving transform of the shift $S_N\mathbf{x}$ is formed by first shifting the columns and then shifting the rows of the 0-th column.

The interleaving transform of higher order shifts can be found by iteration, leading to the following result.

Theorem 3.1 *For $\mathbf{x} \in \mathbb{C}^N$*

$$M\left(S_N^k\mathbf{x}\right) = \begin{bmatrix} S\mathbf{x}_{K-k} \ \cdots \ S\mathbf{x}_{K-1} \ \mathbf{x}_0 \ \cdots \ \mathbf{x}_{K-k-1} \end{bmatrix},$$

$$M\left(S_N^K\mathbf{x}\right) = SM\mathbf{x}$$

and

$$M^{k+mK}\mathbf{x} = S^m M\left(S_N^k\mathbf{x}\right), \qquad 0 \le k < K, \ \ 0 \le m < L.$$

Shifts are the building blocks of convolution and correlation. In Chapter 7, the formula in Theorem 3.1 will be generalized to Zak space and in Chapter 8 this generalization is used to give a Zak space realization of convolution and correlation.

4

Finite Fourier Transform

$F = F(N)$, $S = S_N$, $R = R_N$ and $D = D_N$.

The finite Fourier transform matrix

$$F = [w^{mn}]_{0 \le m,n < N} \qquad w - e^{2\pi i \frac{1}{N}},$$

is a highly structured matrix. This structure is expressed in the algebraic properties of F, by the relationships between the columns of F and by formulas relating F with S, R and D. In this section we describe this structure and, perhaps, the main consequence of this structure: the Cooley-Tukey fast Fourier transform algorithm (CT FFT) [13, 40]. As a special case of the CT FFT we derive a formula for the finite Fourier transform of a zero-padded vector. This result will be used in Chapter 14 to describe the Zak transform of a zero-padded vector, an important result for applications. The CT FFT is usually viewed as an algorithm for fast computation. In this and subsequent chapters we extend this role to that of a conceptual tool for studying the Zak transform and its application to echo analysis, correlation and signal design.

The following result is the main tool for investigating the properties of F. In physical terms it describes the constructive interference and reinforcement properties of the finite Fourier transform.

Theorem 4.1

$$\sum_{n=0}^{N-1} w^{mn} = \begin{cases} N, & m = 0, \\ 0, & m \ne 0, \end{cases} \qquad 0 \le m < N.$$

Proof If $m = 0$, the result is clear. Suppose $0 < m < N$. Then $w^m \ne 1$, but as $w^N = 1$,

$$w^m \sum_{n=0}^{N-1} w^{mn} = \sum_{n=0}^{N-1} w^{m(n+1)} = \sum_{n=0}^{N-1} w^{mn}.$$

This can only happen if

M. An et al., *Ideal Sequence Design in Time-Frequency Space*,
DOI 10.1007/978-0-8176-4738-4_4,
© Birkhäuser Boston, a part of Springer Science+Business Media, LLC 2009

$$\sum_{n=0}^{N-1} w^{mn} = 0,$$

completing the proof of the theorem.

Write F in terms of its column vectors,

$$F = [F_0 \;\cdots\; F_{N-1}],$$

where

$$F_m = D^m \mathbf{1} = [w^{mn}]_{0 \le n < N}, \qquad 0 \le m < N.$$

Because F is symmetric,

$$F = \begin{bmatrix} F_0{}^T \\ \vdots \\ F_{N-1}{}^T \end{bmatrix}.$$

Example 4.1 Set $w = e^{2\pi i \frac{1}{5}}$.

$$F(5) = \begin{bmatrix} 1 & 1 & 1 & 1 & 1 \\ 1 & w & w^2 & w^3 & w^4 \\ 1 & w^2 & w^4 & w^6 & w^8 \\ 1 & w^3 & w^6 & w^9 & w^{12} \\ 1 & w^4 & w^8 & w^{12} & w^{16} \end{bmatrix} = \begin{bmatrix} 1 & 1 & 1 & 1 & 1 \\ 1 & w & w^2 & w^3 & w^4 \\ 1 & w^2 & w^4 & w & w^3 \\ 1 & w^3 & w & w^4 & w^2 \\ 1 & w^4 & w^3 & w^2 & w \end{bmatrix}.$$

We see that

$$F_0 = \mathbf{1}, \qquad F_1{}^* = F_4, \qquad F_2{}^* = F_3,$$

and the componentwise products of the column vectors are given by

$$F_0{}^2 = F_0, \qquad F_1{}^2 = F_2, \qquad F_1 F_2 = F_3, \qquad F_1 F_3 = F_4, \qquad F_1 F_4 = F_0,$$

$$F_2{}^2 = F_4, \qquad F_2 F_3 = F_0, \qquad F_2 F_4 = F_1,$$

and

$$F_3{}^2 = F_1, \qquad F_3 F_4 = F_2.$$

If N is even,

$$F_{\frac{N}{2}} = \begin{bmatrix} 1 \\ -1 \\ \vdots \\ 1 \\ -1 \end{bmatrix}.$$

In general we have

$$F_k{}^* = F_{N-k}, \qquad 0 \le k < N,$$

and the componentwise products of the column vectors of F satisfy

$$D^j F_k = F_j F_k = F_{j+k}, \qquad j+k \text{ taken modulo } N.$$

This proves the following result.

Theorem 4.2 $F^* = RF = FR$ *and*

$$D^j F = F S^j, \qquad 0 \le j < N.$$

By Theorem 4.1,

$$F_j{}^T F_k{}^* = \sum_{n=0}^{N-1} w^{(j-k)n} = \begin{cases} N, \ j = k, \\ 0, \ j \ne k, \end{cases} \qquad 0 \le j, k < N,$$

implying the following result.

Theorem 4.3 $FF^* = NI$ *and* $F^{-1} = \frac{1}{N}F^*$.

By Theorems 4.2 and 4.3 we have the following two corollaries.

Corollary 4.1 $F^2 = NR$ *and* $F^4 = N^2 I$.

Proof

$$F^2 = FF^* R = NR$$

and

$$F^4 = N^2 R^2 = N^2 I,$$

completing the proof.

Corollary 4.2 $F^{-1} S^j F = D^{-j}, 0 \le j < N$.

Proof

$$F^{-1} S^j F = F^{-2} D^j F^2 = R D^j R = D^{-j},$$

completing the proof.

The formula from Theorem 4.2,

$$F S^j F^{-1} = D^j, \qquad 0 \le j < N,$$

is especially important. For $\mathbf{x} \in \mathbb{C}^N$, define

$$C(\mathbf{x}) = \sum_{j=0}^{N-1} x_j S^j.$$

Example 4.2 $N = 4$ and $\mathbf{x} \in \mathbb{C}^4$.

$$C(\mathbf{x}) = \begin{bmatrix} x_0 & x_3 & x_2 & x_1 \\ x_1 & x_0 & x_3 & x_2 \\ x_2 & x_1 & x_0 & x_3 \\ x_3 & x_2 & x_1 & x_0 \end{bmatrix}.$$

$C(\mathbf{x})$ is formed by placing $\mathbf{x}$ in the 0-th column and placing shifts of $\mathbf{x}$ in subsequent columns,

$$C(\mathbf{x}) = \begin{bmatrix} \mathbf{x} & S\mathbf{x} & \cdots & S^{N-1}\mathbf{x} \end{bmatrix} = \begin{bmatrix} x_0 & x_{N-1} & \cdots & x_1 \\ x_1 & x_0 & & x_2 \\ & & \cdot & \\ \cdot & & & \cdot \\ & & & \cdot \\ x_{N-1} & x_{N-2} & \cdot & x_0 \end{bmatrix}.$$

We call $C(\mathbf{x})$ the *circulant matrix* defined by $\mathbf{x}$.

The sum and product of two circulant matrices are circulant matrices,

$$C(\mathbf{x}) + C(\mathbf{y}) = C(\mathbf{x} + \mathbf{y}),$$

$$C(\mathbf{x})C(\mathbf{y}) = C(\mathbf{z}) = C(\mathbf{y})C(\mathbf{x}),$$

where

$$z_k = \sum_{j=0}^{N-1} x_j y_{k-j}, \quad 0 < k < N.$$

In the next chapter we identify $C(\mathbf{x})C(\mathbf{y})$ with the convolution of $\mathbf{x}$ and $\mathbf{y}$. Another way of expressing these results is that the collection of all circulant matrices is a commutative algebra of matrices,

$$FC(\mathbf{x})F^{-1} = \sum_{j=0}^{N-1} x_j D^j = D(\mathbf{y}),$$

where $y_n = \sum_{j=0}^{N-1} x_j w^{nj}$, proving the following result.

Theorem 4.4 *For* $\mathbf{x} \in \mathbb{C}^N$

$$FC(\mathbf{x})F^{-1} = D(F\mathbf{x}).$$

Corollary 4.3

$$F^{-1}C(\mathbf{x})F = D(F^*\mathbf{x}).$$

Proof

$$F^{-1}C(\mathbf{x})F = F^{-2}D(F\mathbf{x})F^2 = RD(F\mathbf{x})R^{-1} = D(F^*\mathbf{x}),$$

completing the proof.

$C(\mathbf{x})$ is invertible if and only if $D(F\mathbf{x})$ is invertible. This is the case if all the components of $F\mathbf{x}$ are nonzero, in which case

$$C(\mathbf{x})^{-1} = F^{-1}D(F\mathbf{x})^{-1}F.$$

In particular, the inverse of an invertible circulant matrix is a circulant matrix.

4.1 CT FFT

$N = LK$, where $L > 1$ and $K > 1$ are integers. $D = D_N$, $F = F(N)$, $S = S_N$.

The CT FFT algorithm can be described by a matrix factorization, each factor representing a stage of the algorithm. The goal is not implementation, but rather a formula for the finite Fourier transform of a zero-padded vector.

Define the $K \times K$ diagonal matrix $D_K(N)$ by

$$D_K(N) = D\left([w^k]_{0 \leq k < K}\right), \qquad w = e^{2\pi i \frac{1}{N}}.$$

$D_K(N)$ is the matrix formed from the first K columns and K rows of D.

Example 4.3

$$D_2(6) = \begin{bmatrix} 1 & 0 \\ 0 & w \end{bmatrix}, \qquad w = e^{2\pi i \frac{1}{6}}.$$

Example 4.4

$$D_3(9) = \begin{bmatrix} 1 & & \\ & w & \\ & & w^2 \end{bmatrix}, \qquad w = e^{2\pi i \frac{1}{9}}.$$

$D_K(N)$ does not behave as well as D_K with respect to $F(K)$ and S_K. The basic reason is that $w^K \neq 1$, when $w = e^{2\pi i \frac{1}{N}}$ and $K > 1$. The resulting loss of periodicity of the diagonal entries significantly affects its properties.

D and $D_K(N)$ are related by the following tensor product formula.

Theorem 4.5
$$D = D_L \otimes D_K(N).$$

Proof

$$D_L \otimes D_K(N) = \begin{bmatrix} D_K(N) & & & & \\ & vD_K(N) & & & \\ & & \cdot & & \\ & & & \cdot & \\ & & & & \cdot \, v^{L-1}D_K(N) \end{bmatrix}, \qquad v = e^{2\pi i \frac{1}{L}}.$$

Because

$$vD_K(N) = \begin{bmatrix} v & & & & \\ & vw & & & \\ & & \cdot & & \\ & & & \cdot & \\ & & & & vw^{K-1} \end{bmatrix}$$

and $w^K = v$,

$$vD_K(N) = \begin{bmatrix} w^K & & & & \\ & w^{K+1} & & & \\ & & \cdot & & \\ & & & \cdot & \\ & & & & \cdot \\ & & & & vw^{2K-1} \end{bmatrix}.$$

Continuing in this way, the theorem follows.

Example 4.5

$$D_3 \otimes D_2(6) = \begin{bmatrix} 1 & & & & & \\ & w & & & & \\ & & v & & & \\ & & & vw & & \\ & & & & v^2 & \\ & & & & & v^2w \end{bmatrix} = \begin{bmatrix} 1 & & & & & \\ & w & & & & \\ & & w^2 & & & \\ & & & w^3 & & \\ & & & & w^4 & \\ & & & & & w^5 \end{bmatrix}.$$

Because

$$S_K^{-1} = F(K)D_K F(K)^{-1}$$

the $K \times K$ matrix $C_K(N)$

$$C_K(N) = F(K)D_K(N)F(K)^{-1}$$

is the circulant matrix defined by

$$\frac{1}{K}F(K)\left([w^k]_{0 \le k < K}\right), \qquad w = e^{2\pi i \frac{1}{N}}.$$

Its structure is significantly more complex than that of the shift S_K^{-1}. This results from the lack of periodicity of the vector formed from the diagonal entries of $D_K(N)$.

The matrix $C_K(N)$ plays an important role in the CT FFT and in the formula, derived below, for the finite Fourier transform of zero-padded vectors.

Form the powers

$$C_K(N)^l = F(K)D_K(N)^l F(K)^{-1}, \qquad 0 \le l < L,$$

and define the $N \times K$ matrix

$$\mathcal{C}_K(N) = \begin{bmatrix} I_K \\ C_K(N) \\ \vdots \\ C_K(N)^{L-1} \end{bmatrix}.$$

Define the $N \times N$ diagonal matrix $T_K(N)$ by

$$T_K(N) = \bigoplus_{l=0}^{L-1} D_K(N)^l,$$

where $\bigoplus$ is the matrix direct sum. K and L can be interchanged by stride permutations

$$T_L(N) = P(N, K)T_K(N)P(N, K)^{-1}.$$

Example 4.6

$$T_2(6) = \begin{bmatrix} I_2 & & \\ & D_2(6) & \\ & & D_2(6)^2 \end{bmatrix} = \begin{bmatrix} 1 & & & & & \\ & 1 & & & & \\ & & 1 & & & \\ & & & w & & \\ & & & & 1 & \\ & & & & & w^2 \end{bmatrix}.$$

One form of the CT FFT algorithm is given by the factorization

$$F - P(N, K) \left(I_L \otimes F(K) \right) T_K(N) \left(F(L) \otimes I_K \right).$$

Because

$$\left(I_L \otimes F(K) \right) T_K(N) = \bigoplus_{l=0}^{L-1} F(K)D_K(N)^l = \left(\bigoplus_{l=0}^{L-1} C_K(N)^l \right) \left(I_L \otimes F(K) \right),$$

we can write

$$F = P(N, K) \left(\bigoplus_{l=0}^{L-1} C_K(N)^l \right) \left(F(L) \otimes F(K) \right).$$

4.2 FT of Zero-Padded Vectors

Suppose $\mathbf{x} \in \mathbb{C}^K$ and define $\mathbf{x}^N \in \mathbb{C}^N$ by

$$\mathbf{x}^N = \mathbf{e}_0^L \otimes \mathbf{x} = \begin{bmatrix} \mathbf{x} \\ \mathbf{0} \\ \vdots \\ \mathbf{0} \end{bmatrix}, \qquad \mathbf{0} = \mathbf{0}^K.$$

We call $\mathbf{x}^N$ the *zero-padded vector* in $\mathbb{C}^N$ defined by $\mathbf{x}$.

We can compute

$$P(N, L)F\mathbf{x}^N$$

by the following steps:

$$\left(F(L) \otimes F(K) \right) \mathbf{x}^N = \left(\mathbf{1}^L \otimes F(K) \right) \mathbf{x}$$

and

$$\left(\bigoplus_{l=0}^{L-1} C_K(N)^l \right) (\mathbf{1} \otimes (F(K)\mathbf{x})) = C_K(N)F(K)\mathbf{x},$$

proving the following result.

Theorem 4.6

$$P(N, L)F\mathbf{x}^N = C_K(N)F(K)\mathbf{x}.$$

By Theorem 4.6 the first K components of $P(N, K)F\mathbf{x}^N$ form the vector

$$F(K)\mathbf{x}.$$

The next segment of K components of $P(N, K)F\mathbf{x}^N$ forms the vector

$$C_K(N)F(K)\mathbf{x}.$$

Since $C_K(N)$ is a circulant matrix we can view this vector as a filtered version of $F(K)\mathbf{x}$. Continuing, $P(N, K)F\mathbf{x}^N$ can be viewed as L segments, the l-th segment forming the vector

$$C_K(N)^l F(K)\mathbf{x}, \qquad 0 \le l < L.$$

Example 4.7 Choosing the first column of $F(16)$,

$$\mathbf{x} = \left[w^k \right]_{0 \le k < 16}, \qquad w^{2\pi i \frac{1}{16}},$$

we have

$$F(16)\mathbf{x} = \mathbf{e}_{15}.$$

Figures 4.1 and 4.2 display $F(32)\mathbf{x}^{32}$, $P(32, 2)F(32)\mathbf{x}^{32}$, $F(48)\mathbf{x}^{48}$ and $P(48, 3)F(48)\mathbf{x}^{48}$. The first row of plots is generated using the following MATLAB commands. Note that MATLAB convention for FFT is different from the definition in this text.

```
K=16; R=2;
F=K*ifft(eye(K,K));        % unscaled FT matrix
x=F(:,2);                  % first column
xR=[x;zeros((R-1)*size(x,1),1)];
FxR=ifft(xR);              % modify MATLAB fft convention
figure(1)
plot(real(FxR),'LineWidth',2))
figure(2)
plot(imag(FxR),'LineWidth',2)
```

Example 4.8 For $L = 64$, $\mathbf{x}$ is plotted in Figure 4.2. $\mathbf{x}$ is a shifted, sampled Gaussian and up to a scale factor, its Fourier transform is arbitrarily close to itself. Figure 4.4 displays $F(192)\mathbf{x}^{192}$ and $P(192, 3)F(192)\mathbf{x}^{192}$.

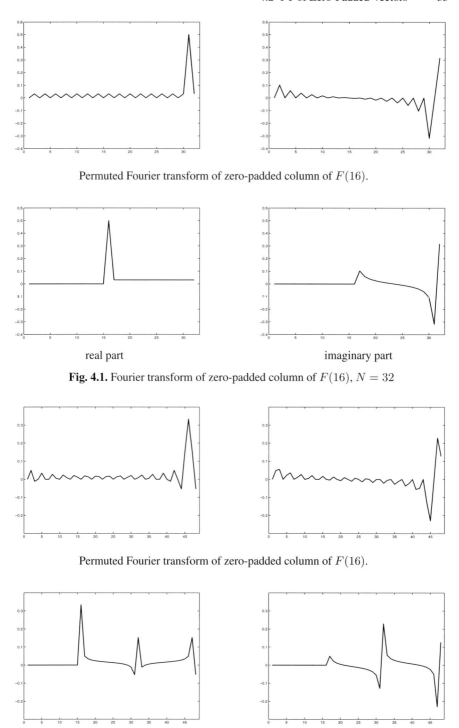

Permuted Fourier transform of zero-padded column of $F(16)$.

real part   imaginary part

Fig. 4.1. Fourier transform of zero-padded column of $F(16)$, $N = 32$

Permuted Fourier transform of zero-padded column of $F(16)$.

real part   imaginary part

Fig. 4.2. Fourier transform of zero-padded column of $F(16)$, $N = 48$

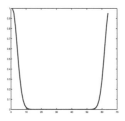

Fig. 4.3. Shifted, sampled Gaussian

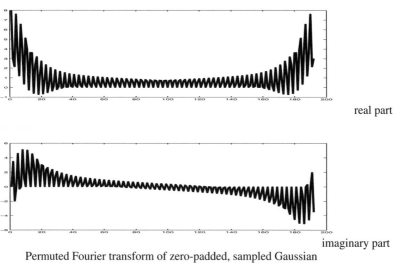

real part

imaginary part

Permuted Fourier transform of zero-padded, sampled Gaussian

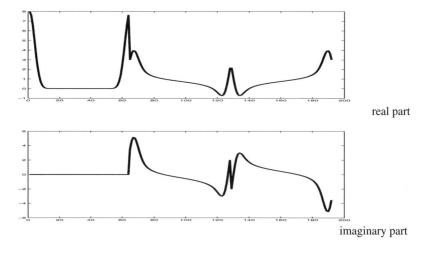

real part

imaginary part

Fig. 4.4. Fourier transform of zero-padded, sampled Gaussian

5

Convolution and Correlation

Shifts are the building blocks for convolution and correlation. Diagonalization of shifts by the Fourier transform leads to algorithms for fast computation of these operations. In Chapter 8 we use the Zak space realization of shifts, discussed in Chapter 7, to derive the Zak space correlation formula. This formula is the main tool for analyzing correlation and designing sequence sets with good correlation properties.

Because radar and sonar signal design is the main purpose of this text, the importance of the shift needs no discussion. Echoes are linear combinations of shifts. However, many of the results, with slight modifications, continue to hold if shifts are replaced by other classes of unitary operations. An interesting modification is presented in Chapter 16.

All definitions and results, unless otherwise specified, refer to cyclic or periodic constructions. We identify $\mathbb{C}^N$ with the space of periodic complex sequences of period N. Throughout this chapter $\mathbf{x}$ and $\mathbf{y}$ are vectors in $\mathbb{C}^N$, $F = F(N)$, $R = R_N$, $S = S_N$ and $D = D_N$.

5.1 Convolution

The *convolution* $\mathbf{u} = \mathbf{x} * \mathbf{y} \in \mathbb{C}^N$ is defined by

$$u_m = \sum_{n=0}^{N-1} x_n y_{m-n}, \qquad 0 \leq m < N,$$

where $m - n$ is taken modulo N. Convolution can be written as a linear combination of shifts of $\mathbf{y}$,

$$\mathbf{x} * \mathbf{y} = \sum_{n=0}^{N-1} x_n S^n \mathbf{y} = C(\mathbf{x}) \mathbf{y}.$$

$C(\mathbf{x})$ is the circulant matrix

$$C(\mathbf{x}) = \sum_{n=0}^{N-1} x_n S^n.$$

M. An et al., *Ideal Sequence Design in Time-Frequency Space*,
DOI 10.1007/978-0-8176-4738-4_5,

By Theorem 4.4

$$Fu = FC(\mathbf{x})F^{-1}F\mathbf{y} = D(F\mathbf{x})F\mathbf{y} = (F\mathbf{x})(F\mathbf{y}),$$

where $(F\mathbf{x})(F\mathbf{y})$ is the componentwise product, proving the following result.

Theorem 5.1

$$\mathbf{x} * \mathbf{y} = C(\mathbf{x})\mathbf{y},$$

and

$$F(\mathbf{x} * \mathbf{y}) = (F\mathbf{x})(F\mathbf{y}),$$

where $(F\mathbf{x})(F\mathbf{y})$ is the componentwise product.

In detail,

$$\mathbf{x} * \mathbf{y} = \begin{bmatrix} x_0 & x_{N-1} & \cdots & x_1 \\ x_1 & x_0 & & x_2 \\ & & & \\ \cdot & \cdot & & \cdot \\ & & & \\ x_{N-1} & x_{N-2} & \cdot & x_0 \end{bmatrix} \mathbf{y}.$$

Example 5.1 $v = e^{2\pi i \frac{1}{3}}$. Using $1 + v + v^2 = 0$, we have

$$\begin{bmatrix} 1 \\ 0 \\ 1 \end{bmatrix} * \begin{bmatrix} a \\ b \\ c \end{bmatrix} = \begin{bmatrix} 1 & 1 & 0 \\ 0 & 1 & 1 \\ 1 & 0 & 1 \end{bmatrix} \begin{bmatrix} a \\ b \\ c \end{bmatrix} = \begin{bmatrix} a+b \\ b+c \\ a+c \end{bmatrix},$$

$$F \begin{bmatrix} a+b \\ b+c \\ a+c \end{bmatrix} = \begin{bmatrix} 2(a+b+c) \\ -(c+va+v^2b) \\ -(c+vb+v^2a) \end{bmatrix},$$

and

$$\left(F \begin{bmatrix} 1 \\ 0 \\ 1 \end{bmatrix} \right) \left(F \begin{bmatrix} a \\ b \\ c \end{bmatrix} \right) = \begin{bmatrix} 2 \\ 1+v^2 \\ 1+v \end{bmatrix} \begin{bmatrix} a+b+c \\ a+vb+v^2c \\ a+v^2b+vc \end{bmatrix}$$

$$= \begin{bmatrix} 2(a+b+c) \\ -(c+va+v^2b) \\ -(c+vb+v^2a) \end{bmatrix}.$$

Example 5.2

$$\begin{bmatrix} 1 \\ i \\ -1 \\ -i \end{bmatrix} * \begin{bmatrix} 1 \\ i \\ -1 \\ -i \end{bmatrix} = \begin{bmatrix} 1 & -i & -1 & i \\ i & 1 & -i & -1 \\ -1 & i & 1 & -i \\ -i & -1 & i & 1 \end{bmatrix} \begin{bmatrix} 1 \\ i \\ -1 \\ -i \end{bmatrix} = 4 \begin{bmatrix} 1 \\ i \\ -1 \\ -i \end{bmatrix},$$

$$4F\begin{bmatrix}1\\i\\-1\\-i\end{bmatrix}=\begin{bmatrix}0\\0\\0\\16\end{bmatrix},$$

and

$$\left(F\begin{bmatrix}1\\i\\-1\\-i\end{bmatrix}\right)\left(F\begin{bmatrix}1\\i\\-1\\-i\end{bmatrix}\right)=\begin{bmatrix}0\\0\\0\\16\end{bmatrix}.$$

Example 5.3 Figure 5.1 and Figure 5.2 display plots of **x** and **y**, where **x** is the first column of $F(64)$ and **y** is the shifted Gaussian.

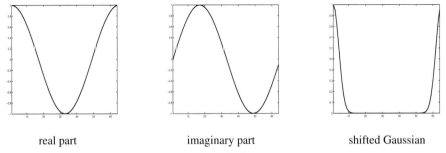

real part   imaginary part   shifted Gaussian

Fig. 5.1. First column of $F(64)$ and shifted Gaussian

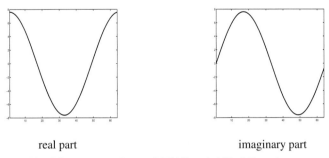

real part   imaginary part

Fig. 5.2. **x** ∗ **y**, a column of $F(64)$ and shifted Gaussian

Example 5.4 Define $\mathbf{x}\in\mathbb{C}^{256}$ by

$$x(t)=e^{32\pi i(\frac{t}{256})^2},\qquad 0\le t<256.$$

x is a chirp which will be studied in detail beginning with Chapter 6. Figure 5.3 displays **x**∗**x**.

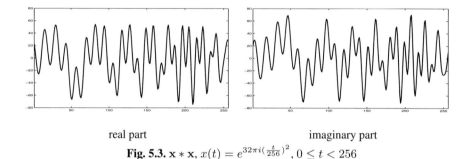

real part   imaginary part

Fig. 5.3. $\mathbf{x} * \mathbf{x}$, $x(t) = e^{32\pi i(\frac{t}{256})^2}$, $0 \leq t < 256$

5.2 Correlation

The *correlation* $\mathbf{v} = \mathbf{x} \circ \mathbf{y} \in \mathbb{C}^N$ is defined by

$$v_m = \sum_{n=0}^{N-1} x_n y_{n-m}^*, \qquad 0 \leq m < N,$$

where $n - m$ is taken modulo N. $\mathbf{x} \circ \mathbf{x}$ is called the *autocorrelation* of $\mathbf{x}$, while if $\mathbf{x}$ and $\mathbf{y}$ are distinct, $\mathbf{x} \circ \mathbf{y}$ is called the *cross correlation* of $\mathbf{x}$ and $\mathbf{y}$.

The correlation can be written using inner products,

$$v_m = \langle \mathbf{x}, S^m \mathbf{y} \rangle, \qquad 0 \leq m < N.$$

Convolution and correlation are related by the formula

$$\mathbf{x} \circ \mathbf{y} = \mathbf{x} * (R\mathbf{y}^*).$$

By Theorem 5.1

$$\mathbf{x} \circ \mathbf{y} = C(\mathbf{x}) R\mathbf{y}^*$$

and

$$F(\mathbf{x} \circ \mathbf{y}) = FC(\mathbf{x}) F^{-1}(FR\mathbf{y}^*) = D(F\mathbf{x})(F\mathbf{y})^*,$$

proving the following result.

Theorem 5.2

$$\mathbf{x} \circ \mathbf{y} = C(\mathbf{x}) R\mathbf{y}^*, \quad and \quad F(\mathbf{x} \circ \mathbf{y}) = (F\mathbf{x})(F\mathbf{y})^*,$$

where $(F\mathbf{x})(F\mathbf{y})^*$ *is the componentwise product.*

In detail,

$$\mathbf{x} \circ \mathbf{y} = \begin{bmatrix} x_0 & x_{N-1} & \cdots & x_1 \\ x_1 & x_0 & & x_2 \\ & & & \\ & \cdot & \cdot & \cdot \\ & & & \\ x_{N-1} & x_{N-2} & \cdot & x_0 \end{bmatrix} \begin{bmatrix} y_0^* \\ y_{N-1}^* \\ \cdot \\ \cdot \\ \cdot \\ y_1^* \end{bmatrix}.$$

Example 5.5 $v = e^{2\pi i \frac{1}{3}}$. Using $1 + v + v^2 = 0$, we have

$$\begin{bmatrix} 1 \\ 0 \\ 1 \end{bmatrix} \circ \begin{bmatrix} a \\ b \\ c \end{bmatrix} = \begin{bmatrix} 1 & 1 & 0 \\ 0 & 1 & 1 \\ 1 & 0 & 1 \end{bmatrix} \begin{bmatrix} a^* \\ c^* \\ b^* \end{bmatrix} = \begin{bmatrix} a^* + c^* \\ b^* + c^* \\ a^* + b^* \end{bmatrix},$$

$$F \begin{bmatrix} a^* + c^* \\ b^* + c^* \\ a^* + b^* \end{bmatrix} = \begin{bmatrix} 2(a^* + b^* + c^*) \\ -(b^* + va^* + v^2 c^*) \\ -(b^* + vc^* + v^2 a^*) \end{bmatrix},$$

and

$$\left(F \begin{bmatrix} 1 \\ 0 \\ 1 \end{bmatrix} \right) \left(F \begin{bmatrix} a \\ b \\ c \end{bmatrix} \right)^* = \begin{bmatrix} 2 \\ -v \\ -v^2 \end{bmatrix} \begin{bmatrix} a^* + b^* + c^* \\ a^* + vc^* + v^2 b^* \\ a^* + vb^* + v^2 c^* \end{bmatrix}$$

$$= \begin{bmatrix} 2(a^* + b^* + c^*) \\ -(b^* + va^* + v^2 c^*) \\ -(b^* + vc^* + v^2 a^*) \end{bmatrix}.$$

Example 5.6 Figure 5.4 displays the correlation of the first column of $F(64)$ and the shifted Gaussian.

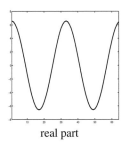   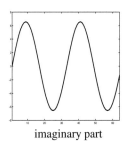

real part   imaginary part

Fig. 5.4. $\mathbf{x} \circ \mathbf{y}$, a column of $F(64)$ and Gaussian

Example 5.7 Continuing with Example 5.4, the autocorrelation of a discrete chirp is displayed in Figure 5.7.

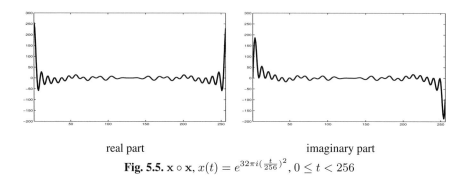

real part   imaginary part

Fig. 5.5. $\mathbf{x} \circ \mathbf{x}$, $x(t) = e^{32\pi i (\frac{t}{256})^2}$, $0 \le t < 256$

A vector $\mathbf{x} \in \mathbb{C}^N$ satisfies the *ideal autocorrelation property* if

$$\mathbf{x} \circ \mathbf{x} = ||\mathbf{x}||^2 \mathbf{e}_0.$$

Set $\mathbf{u} = \mathbf{x} \circ \mathbf{x}$. Because

$$\langle S^r \mathbf{x}, S^s \mathbf{x} \rangle = \langle \mathbf{x}, S^{s-r} \mathbf{x} \rangle = u_{s-r}, \qquad 0 \le r, s < N,$$

$\mathbf{x}$ satisfies the ideal autocorrelation property if and only if the collection of shifts of $\mathbf{x}$

$$\{S^r \mathbf{x} : 0 \le r < N\}$$

is an orthogonal basis of $\mathbb{C}^N$.

Theorem 5.3 *If* $||\mathbf{x}|| = 1$, *then* $\mathbf{x}$ *satisfies the ideal autocorrelation property if and only if the collection of shifts of* $\mathbf{x}$ *forms an orthonormal basis of* $\mathbb{C}^N$.

Set $\mathbf{v} = \mathbf{x} \circ \mathbf{y}$. If $\mathbf{x}$ and $\mathbf{y}$ satisfy the ideal autocorrelation property, then we must have

$$|v_n| \ge \frac{||\mathbf{x}|| \, ||\mathbf{y}||}{\sqrt{N}}, \qquad 0 \le n < N.$$

This bound is called the *Sarwate bound* [32]. The vector pair $(\mathbf{x}, \mathbf{y})$ satisfies the *ideal correlation property* if $\mathbf{x}$ and $\mathbf{y}$ satisfy the ideal autocorrelation property and the components of the cross correlation $\mathbf{v}$ satisfy

$$|v_n| = \frac{||\mathbf{x}|| \, ||\mathbf{y}||}{\sqrt{N}}, \qquad 0 \le n < N.$$

Example 5.8 Set $F = F(5)$ and

$$\mathbf{x} = \begin{bmatrix} F_0 \\ F_1 \\ F_2 \\ F_3 \\ F_4 \end{bmatrix}, \quad \mathbf{y} = (S_5 \otimes I_5)\,\mathbf{x} = \begin{bmatrix} F_1 \\ F_2 \\ F_3 \\ F_4 \\ F_0 \end{bmatrix}, \quad \mathbf{z} = (\mu_2 \otimes I_5)\,\mathbf{x} = \begin{bmatrix} F_0 \\ F_2 \\ F_4 \\ F_1 \\ F_3 \end{bmatrix}.$$

Direct computation shows that $\mathbf{x}$, $\mathbf{y}$ and $\mathbf{z}$ all satisfy the ideal autocorrelation property. Figure 5.6 displays cross correlations of $\mathbf{x}$, $\mathbf{y}$ and $\mathbf{z}$.

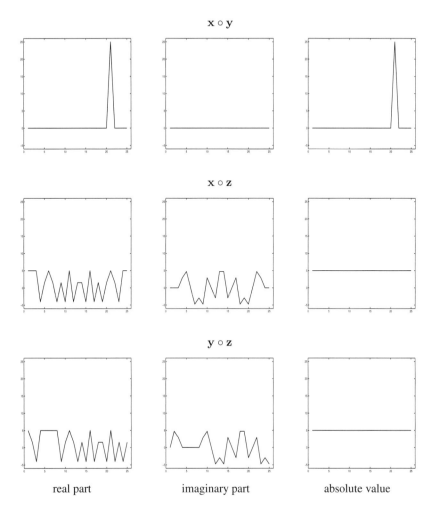

Fig. 5.6. Cross correlations

5.3 Characterization of Ideal Correlation

In this section we describe an algorithm for computing vector pairs in $\mathbb{C}^N$ satisfying ideal correlation. The algorithm requires finding $\mathbf{w} \in \mathbb{C}_1^N$ such that

$$\frac{1}{\sqrt{N}}F\mathbf{w} \in \mathbb{C}_1^N.$$

Each $\mathbf{w} \in \mathbb{C}_1^N$ satisfying this condition leads to a large collection of vectors $\mathbf{a}$, $\mathbf{b} \in \mathbb{C}^N$ such that $(\mathbf{a}, \mathbf{b})$ satisfies ideal correlation. However, finding such a $\mathbf{w}$ is itself a difficult problem and the algorithm is not especially useful. Moreover, the components of vectors $\mathbf{a}$ and $\mathbf{b}$ will not generally have constant absolute value.

Suppose $\mathbf{a} \in \mathbb{C}^N$ satisfies *normalized ideal autocorrelation,*

$$\mathbf{a} \circ \mathbf{a} = \mathbf{e}_0.$$

Then

$$F(\mathbf{a} \circ \mathbf{a}) = (F\mathbf{a})(F\mathbf{a})^* = 1$$

and the components of $F\mathbf{a}$ have absolute value one, proving the following result.

Theorem 5.4 $\mathbf{a} \in \mathbb{C}^N$ *satisfies normalized ideal autocorrelation if and only if*

$$\mathbf{a} = F^{-1}\mathbf{u}, \quad \mathbf{u} \in \mathbb{C}_1^N.$$

Suppose $\mathbf{a}, \mathbf{b} \in \mathbb{C}^N$ satisfy normalized ideal correlation,

$$\mathbf{a} = F^{-1}\mathbf{u} \text{ and } \mathbf{b} = F^{-1}\mathbf{v}, \quad \mathbf{u}, \mathbf{v} \in \mathbb{C}_1^N,$$

and $\mathbf{c} = \mathbf{a} \circ \mathbf{b}$ satisfies

$$|c_m| = \frac{1}{\sqrt{N}}, \quad 0 \le m < N.$$

Set $\mathbf{w} = F\mathbf{c}$.

$$\mathbf{w} = (F\mathbf{a})(F\mathbf{b})^* = \mathbf{u}\mathbf{v}^* \in \mathbb{C}_1^N.$$

Because

$$\frac{1}{\sqrt{N}}F\mathbf{w} = \frac{1}{\sqrt{N}}F^2\mathbf{c} = \sqrt{N}R\mathbf{c},$$

we have

$$\frac{1}{\sqrt{N}}F\mathbf{w} \in \mathbb{C}_1^N.$$

The converse is proved by reversing the steps, proving the following result.

Theorem 5.5 *Suppose* $\mathbf{a}$ *and* $\mathbf{b}$ *in* $\mathbb{C}^N$ *satisfy normalized ideal autocorrelation. Then* $(\mathbf{a}, \mathbf{b})$ *satisfies ideal correlation if and only if*

$$\mathbf{w} = (F\mathbf{a})(F\mathbf{b})^*$$

satisfies

$$\frac{1}{\sqrt{N}}F\mathbf{w} \in \mathbb{C}_1^N.$$

Algorithm 1 *Compute* $\mathbf{a}$ *and* $\mathbf{b}$ *in* $\mathbb{C}^N$ *satisfying normalized ideal correlation.*

- *Choose* $\mathbf{w} \in \mathbb{C}_1^N$ *such that* $\frac{1}{\sqrt{N}}F\mathbf{w} \in \mathbb{C}_1^N$.

- *Choose* **u** *and* **v** *in* $\mathbb{C}_1^N$ *such that*

$$\mathbf{w} = \mathbf{u}\mathbf{v}^*.$$

- *Compute*

$$\mathbf{a} = F^{-1}\mathbf{u} \quad and \quad \mathbf{b} = F^{-1}\mathbf{v}.$$

The following examples provide **w** satisfying the conditions of Algorithm 1.

Example 5.9 $N = 2$.

$$\mathbf{w} = \begin{bmatrix} 1 \\ i \end{bmatrix} \quad \text{and} \quad \frac{1}{\sqrt{2}}F(2)\mathbf{w} = \frac{1}{\sqrt{2}}\begin{bmatrix} 1+i \\ 1-i \end{bmatrix}.$$

Example 5.10 $N = 3$ and $\rho = e^{2\pi i \frac{1}{3}}$.

$$\mathbf{w} = \begin{bmatrix} 1 \\ 1 \\ \rho \end{bmatrix} \quad \text{and} \quad \frac{1}{\sqrt{3}}F(3)\mathbf{w} = \frac{1}{\sqrt{3}}\begin{bmatrix} 2+\rho \\ 2+\rho \\ 1+2\rho^2 \end{bmatrix}.$$

Example 5.11 $N = 5$ and $\rho = e^{2\pi i \frac{1}{5}}$.

$$\mathbf{w} = \begin{bmatrix} 1 \\ 1 \\ \rho \\ \rho^3 \\ \rho \end{bmatrix} \quad \text{and} \quad \frac{1}{\sqrt{5}}F(5)\mathbf{w} = \frac{1}{\sqrt{5}}\begin{bmatrix} 2+2\rho+\rho^3 \\ 2+2\rho+\rho^3 \\ 2+\rho^2+2\rho^4 \\ 1+2\rho^2+2\rho^3 \\ 2+\rho^2+2\rho^4 \end{bmatrix}.$$

5.4 Acyclic Convolution and Correlation

$L > 1$ and $K > 1$ are integers and $N = L + K - 1$.

In applications it is more usual to use acyclic convolution and acyclic correlation rather than (cyclic) convolution and (cyclic) correlation. We define these terms and show how to embed acyclic convolution in a larger size (cyclic) convolution and acyclic correlation in a larger size (cyclic) correlation. For sufficiently large problems, large defined by target computer, computing acyclic convolution and acyclic correlation by (cyclic) convolution and (cyclic) correlation leads to significantly faster computation due to the Cooley-Tukey fast Fourier transform algorithm. However, the main purpose of this embedding is that it places acyclic convolution and acyclic correlation inside the rich algebraic structure of the corresponding cyclic structure.

Suppose throughout this section that $\mathbf{x} \in \mathbb{C}^L$ and $\mathbf{y} \in \mathbb{C}^K$ and x and y are sequences,

$$x(n) = \begin{cases} x_n, & 0 \le n < L, \\ 0, & \text{otherwise,} \end{cases} \quad \text{and} \quad y(n) = \begin{cases} y_n, & 0 \le n < K, \\ 0, & \text{otherwise.} \end{cases}$$

Define the sequence z by

$$z(n) = \sum_{l=0}^{L-1} x(l)y(n-l), \qquad n \in \mathbb{Z}.$$

Because $z(n) = 0$ unless $0 \le n < N$, we identify the sequence z with the vector

$$\mathbf{z} = \mathbf{x} *_a \mathbf{y} \in \mathbb{C}^N,$$

and call $\mathbf{x} *_a \mathbf{y}$ the *acyclic convolution* of $\mathbf{x}$ and $\mathbf{y}$.

Example 5.12 $L = 2$, $K = 3$.

$$z_0 = x_0 y_0$$
$$z_1 = x_0 y_1 + x_1 y_0$$
$$z_2 = x_0 y_2 + x_1 y_1$$
$$z_3 = x_1 y_2$$

which we can write as

$$\mathbf{z} = \begin{bmatrix} x_0 & 0 & 0 \\ x_1 & x_0 & 0 \\ 0 & x_1 & x_0 \\ 0 & 0 & x_1 \end{bmatrix} \begin{bmatrix} y_0 \\ y_1 \\ y_2 \end{bmatrix}.$$

Example 5.13 $L = 3$, $K = 2$.

$$\mathbf{z} = \begin{bmatrix} x_0 & 0 \\ x_1 & x_0 \\ x_2 & x_1 \\ 0 & x_2 \end{bmatrix} \begin{bmatrix} y_0 \\ y_1 \end{bmatrix}.$$

The acyclic convolution can be written as the (cyclic) convolution of zero-padded vectors. Zero-pad $\mathbf{x}$ and $\mathbf{y}$ to vectors $\mathbf{x}^N$ and $\mathbf{y}^N$ in $\mathbb{C}^N$,

$$\mathbf{x}^N = \begin{bmatrix} \mathbf{x} \\ 0 \\ \vdots \\ 0 \end{bmatrix} \quad \text{and} \quad \mathbf{y}^N = \begin{bmatrix} \mathbf{y} \\ 0 \\ \vdots \\ 0 \end{bmatrix}.$$

The cyclic convolution $\mathbf{w} = \mathbf{x}^N * \mathbf{y}^N$ can be written as

$$w_n = \sum_{l=0}^{N-1} x_l y_{n-l}, \qquad 0 \le n < N,$$

where $n - l$ is taken modulo N. Because

$$x_L = \cdots = x_{N-1} = 0 = y_K \cdots = y_{N-1},$$

we can write

$$w_n = \sum_{l=0}^{L-1} x(l)y(n-l), \qquad 0 \le n < N,$$

without the modulo N condition, proving the following result.

Theorem 5.6

$$\mathbf{x} *_a \mathbf{y} = \mathbf{x}^N * \mathbf{y}^N = C\left(\mathbf{x}^N\right)\mathbf{y}^N.$$

Example 5.14 $L = 2,\ K = 3$.

$$\mathbf{x} *_a \mathbf{y} = \begin{bmatrix} x_0 & 0 & 0 & x_1 \\ x_1 & x_0 & 0 & 0 \\ 0 & x_1 & x_0 & 0 \\ 0 & 0 & x_1 & x_0 \end{bmatrix} \begin{bmatrix} y_0 \\ y_1 \\ y_2 \\ 0 \end{bmatrix}.$$

Compare with Example 5.12.

Example 5.15 $L = 4,\ K = 3$.

$$\mathbf{x} *_a \mathbf{y} = \begin{bmatrix} x_0 & 0 & 0 & x_3 & x_2 & x_1 \\ x_1 & x_0 & 0 & 0 & x_3 & x_2 \\ x_2 & x_1 & x_0 & 0 & 0 & x_3 \\ x_3 & x_2 & x_1 & x_0 & 0 & 0 \\ 0 & x_3 & x_2 & x_1 & x_0 & 0 \\ 0 & 0 & x_3 & x_2 & x_1 & x_0 \end{bmatrix} \begin{bmatrix} y_0 \\ y_1 \\ y_2 \\ 0 \\ 0 \\ 0 \end{bmatrix}.$$

Compare with Example 5.13.

The advantage of replacing the acyclic convolution by the cyclic convolution of zero-padded vectors by the formula in Theorem 5.6 is that the finite Fourier transform of $\mathbf{x} *_a \mathbf{y}$ can be computed by the componentwise product

$$F(N)\left(\mathbf{x} *_a \mathbf{y}\right) = \left(F(N)\mathbf{x}^N\right)\left(F(N)\mathbf{y}^N\right).$$

Computing $\mathbf{x} *_a \mathbf{y}$ by taking the inverse finite Fourier transform requires three finite Fourier transforms. The computational efficiency of the finite Fourier transform for sufficiently large N leads to a fast computation of the acyclic convolution.

Example 5.16 Figure 5.7 compares the cyclic convolution and acyclic convolution of the first column of $F(64)$. Note that the convolution of the column by itself is multiplication by 64.

$$F_1 * F_1$$

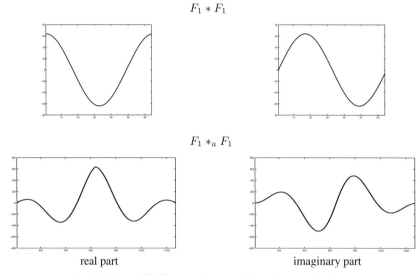

$$F_1 *_a F_1$$

real part   imaginary part

Fig. 5.7. $F = F(64)$, convolution and acyclic convolution

The acyclic correlation is defined in a similar way. Define the sequence u by

$$u(n) = \sum_{l=0}^{L-1} x(l)y^*(l - n + K - 1), \qquad n \in \mathbb{Z}.$$

If $n < 0$, then

$$l - n + K \geq K, \qquad 0 \leq l < L,$$

and $u_n = 0$, and if $n \geq N$, then

$$l - n + K - 1 < 0, \qquad 0 \leq l < L,$$

and $u_n = 0$. We can identify the sequence u with the vector

$$\mathbf{u} = \mathbf{x} \circ_a \mathbf{y} \in \mathbb{C}^N,$$

and call $\mathbf{x} \circ_a \mathbf{y}$ the *acyclic correlation* of $\mathbf{x}$ and $\mathbf{y}$.

Example 5.17 $L = 3, K = 2.$

$$u_0 = x_0 y_2^*,$$
$$u_1 = x_0 y_1^* + x_1 y_2^*,$$
$$u_2 = x_0 y_3^* + x_1 y_1^*,$$
$$u_3 = x_1 y_0^*,$$

which we can write as

$$\mathbf{x} \circ_a \mathbf{y} = \begin{bmatrix} x_0 & 0 & 0 & x_1 \\ x_1 & x_0 & 0 & 0 \\ 0 & x_1 & x_0 & 0 \\ 0 & 0 & x_1 & 0 \end{bmatrix} \begin{bmatrix} y_2^* \\ y_1^* \\ y_0^* \\ 0 \end{bmatrix}.$$

Example 5.18 $L = 4$, $K = 3$.

$$\mathbf{x} \circ_a \mathbf{y} = \begin{bmatrix} x_0 & 0 & 0 & x_3 & x_2 & x_1 \\ x_1 & x_0 & 0 & 0 & x_3 & x_2 \\ x_2 & x_1 & x_0 & 0 & 0 & x_3 \\ x_3 & x_2 & x_1 & x_0 & 0 & 0 \\ 0 & x_3 & x_2 & x_1 & x_0 & 0 \\ 0 & 0 & x_3 & x_2 & x_1 & x_0 \end{bmatrix} \begin{bmatrix} y_2^* \\ y_1^* \\ y_0^* \\ 0 \\ 0 \\ 0 \end{bmatrix}.$$

In general we have the following result.

Theorem 5.7

$$\mathbf{x} \circ_a \mathbf{y} = \mathbf{x}^N * \left(S_K^{-1} R_K \mathbf{y}^* \right)^N = C\left(\mathbf{x}^N \right) \left(S_K^{-1} R_K \mathbf{y}^* \right)^N,$$

where

$$\begin{bmatrix} y_{K-1} \\ \vdots \\ y_0 \end{bmatrix} = S_K^{-1} R_K \mathbf{y}.$$

The acyclic correlation of $\mathbf{x}$ and $\mathbf{y}$ can be written in terms of inner products, but the shift operator on sequences must be used. For a sequence y define the sequence Sy by

$$Sy(n) = y(n-1), \quad n \in \mathbb{Z}.$$

It will always be clear from context whether S is the shift operator on sequences or the shift permutation matrix. Viewing $\mathbf{x} \in \mathbb{C}^L$ and $\mathbf{y} \in \mathbb{C}^K$ as sequences x and y,

$$w_n = \langle x, S^{-(K-1-n)} y \rangle, \quad n \in \mathbb{Z}.$$

Example 5.19 $L = 2$, $K = 3$.

	·	−1	0	1	0	·	·
x	·	·	0	x_0	x_1	0	·
$S^{-2}y$	·	y_0	y_1	y_2	0		· · ·
$S^{-1}y$	·	0	y_0	y_1	y_2	0	· · ·
y		·	0	y_0	y_1	y_2	0 · ·
Sy		·	0	0	y_0	y_1	y_2 0

Example 5.20 Continuing with Example 5.8, Figures 5.8 and 5.9 display the acyclic autocorrelations and acyclic cross correlations. $F = F(5)$ and

$$
\mathbf{x} = \begin{bmatrix} F_0 \\ F_1 \\ F_2 \\ F_3 \\ F_4 \end{bmatrix}, \quad \mathbf{y} = (S_5 \otimes I_5)\,\mathbf{x}, \quad \mathbf{z} = (\mu_2 \otimes I_5)\,\mathbf{x}.
$$

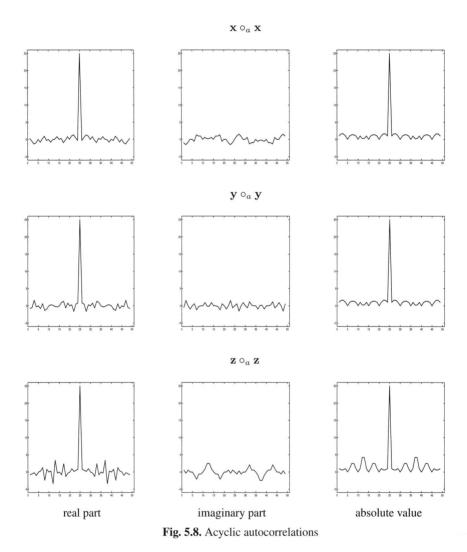

$$\mathbf{x} \circ_a \mathbf{x}$$

$$\mathbf{y} \circ_a \mathbf{y}$$

$$\mathbf{z} \circ_a \mathbf{z}$$

real part   imaginary part   absolute value

Fig. 5.8. Acyclic autocorrelations

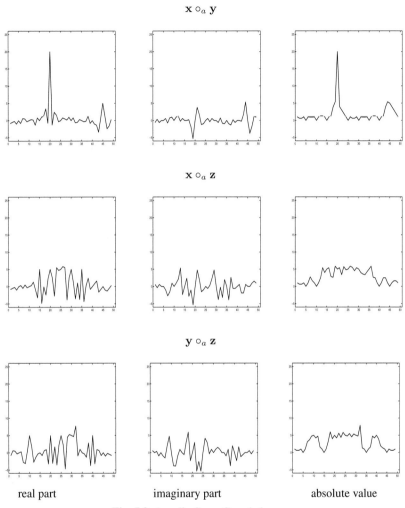

Fig. 5.9. Acyclic Cross Correlations

6

Discrete Chirps

The linear frequency modulated (FM) chirp is one of the most ubiquitous waveforms in digital signal processing with applications including radar, sonar, spread spectrum communications and optics. While the Fourier transform of a discrete chirp is, in general, as complex as the chirp itself, the finite Zak transform of a chirp is sparse and highly structured, thereby permitting generalized sequence design strategies. In this chapter we study the correlation properties of critically sampled and subsampled linear FM chirp pulses.

A *linear FM chirp pulse* is any signal of the form

$$x(t) = \begin{cases} e^{\pi i \gamma t^2} e^{2\pi i \nu t}, & 0 \leq t < T, \\ 0, & \text{otherwise,} \end{cases} \qquad \gamma \neq 0, \ t \in \mathbf{R}.$$

T is the *time duration*, γ is the *chirp rate* and ν is the *carrier frequency* of the chirp pulse. Using the standard definition of bandwidth of a linear FM chirp pulse [29], we obtain $B = \gamma T$ for the *bandwidth* of x. Assume throughout that the time-bandwidth product

$$M = \gamma T^2$$

is a nonzero integer.

The *critically sampled* chirp pulse is the sequence formed by sampling at the points

$$\frac{m}{M} T, \qquad m \in \mathbb{Z}.$$

$$x_s(m) = x\left(\frac{m}{M} T\right) = \begin{cases} e^{\pi i \frac{m^2}{M}} e^{2\pi i f \frac{m}{M}}, & 0 \leq m < M, \\ 0, & \text{otherwise,} \end{cases} \qquad m \in \mathbb{Z},$$

where $f = \nu T$.

For each integer $N > 1$ dividing M we can form a *subsampled* chirp pulse by sampling at the points

$$\frac{n}{N} T, \qquad n \in \mathbb{Z}.$$

M. An et al., *Ideal Sequence Design in Time-Frequency Space*,
DOI 10.1007/978-0-8176-4738-4_6,
© Birkhäuser Boston, a part of Springer Science+Business Media, LLC 2009

$$x_u(n) = x\left(\frac{n}{N}T\right) = \begin{cases} e^{\pi i u \frac{n^2}{N}} e^{2\pi i f \frac{n}{N}}, & 0 \le n < N, \\ 0, & \text{otherwise,} \end{cases} \quad n \in \mathbb{Z},$$

where $u = \frac{M}{N}$ is the *discrete chirp rate* and $f = \nu T$ is the *discrete chirp carrier frequency*.

A *discrete chirp* of period N is any sequence of the form

$$x_u(n) = e^{\pi i u \frac{n^2}{N}} e^{2\pi i f \frac{n}{N}}, \quad n \in \mathbb{Z},$$

where u is a nonzero integer and x_u is periodic mod N,

$$x_u(n+N) = x_u(n), \quad n \in \mathbb{Z}.$$

Periodicity mod N is equivalent to the property $x(N) = 1$. This will be the case if and only if

$$uN + 2f \in 2\mathbb{Z}.$$

In particular, we require $2f \in \mathbb{Z}$. Discrete chirps have been extensively studied in [7, 8, 9, 10]. Several results from these works will be rederived in this chapter.

In [37] we considered different and, in several ways, more general classes of discrete chirps. For the most part the results derived in this work can be extended to these more general classes. However, as a key goal is to use the Zack space representation of discrete chirps as a motivation for sequence design procedures, we chose to limit the discussion to the discrete chirps defined above.

Example 6.1 For $T = 40$, $\gamma = 0.0625$ and $\nu = 0$, the time-bandwidth product $M = \gamma T^2 = 100$. Figures 6.1 and 6.2 display a sampled linear FM chirp pulse. Figure 6.3 displays the critical sample values on the same axis as the linear FM chirp pulse.

real part

imaginary part

Fig. 6.1. Linear FM chirp pulse

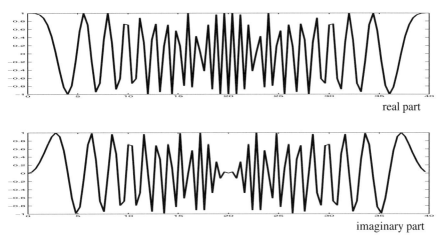

Fig. 6.2. Critically sampled chirp pulse

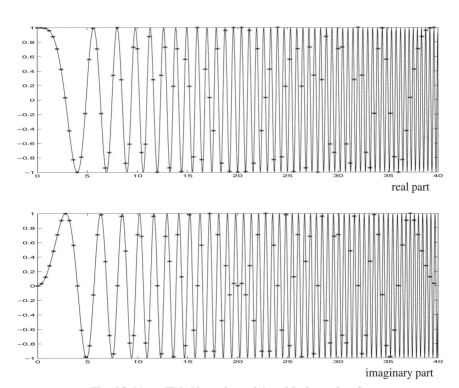

Fig. 6.3. Linear FM chirp pulse and the critical sample values

The following easily derived result captures the essence of a discrete chirp and will be repeatedly used throughout this work.

Theorem 6.1 *If x_u is a discrete chirp, then*

$$x_u(n+m) = x_u(n)x_u(m)e^{2\pi i u \frac{nm}{N}}, \qquad n, m \in \mathbb{Z}.$$

Proof

$$x_u(n+m) = e^{\pi i \frac{(n+m)^2}{N}} e^{2\pi i f \frac{(n+m)}{N}}$$
$$= e^{\pi i u \frac{n^2}{N}} e^{2\pi i f \frac{n}{N}} e^{\pi i u \frac{m^2}{N}} e^{2\pi i f \frac{m}{N}} e^{2\pi i u \frac{nm}{N}}$$
$$= x_u(n)x_u(m)e^{2\pi i u \frac{nm}{N}},$$

completing the proof.

Fourier theory expands sequences of period N in terms of the exponential sequences χ_m, $m \in \mathbb{Z}/N$,

$$\chi_m(n) = e^{2\pi i \frac{mn}{N}}, \qquad n \in \mathbb{Z}.$$

The exponential sequences of period N are characterized by the additive character condition

$$\chi(n_1 + n_2) = \chi(n_1)\chi(n_2), \qquad n_1, n_2 \in \mathbb{Z}.$$

By Theorem 6.1, discrete chirps are perhaps the simplest extensions of exponential sequences in the sense that they satisfy the additive character condition, up to the phase factor $e^{2\pi i u \frac{nm}{N}}$.

The discrete chirp x_u of period N is called a *unit* discrete chirp if

$$(u, N) = 1.$$

We cannot assume $0 < u < N$ because

$$x_{u+N}(n) = e^{\pi i n^2} x_u(n)$$

equals $x_u(n)$ if and only if n is even. In the next chapter, we associate the unit discrete chirp x_u with the unit permutation μ_u.

6.1 Correlation Properties of Discrete Chirps

We will show that the collection of shifts of a unit discrete chirp of period N is, up to a scalar factor, an orthonormal basis of the space of sequences of period N. As a result, the Fourier expansion of sequences of period N in terms of the exponential sequences can be replaced by expansions in terms of shifts of unit discrete chirps of period N.

Under the usual identification between $\mathbb{C}^N$ and the space of sequences of period N, we identify vectors $\mathbf{x} \in \mathbb{C}^N$ with sequences x of period N.

Theorem 6.2 *If $\mathbf{x}_u$ is a unit discrete chirp in $\mathbb{C}^N$, then $\mathbf{x}_u$ satisfies ideal autocorrelation.*

Proof By Theorem 6.1,

$$(\mathbf{x}_u \circ \mathbf{x}_u)(k) = \sum_{n=0}^{N-1} x_u(n)x_u^*(n-k) = x_u^*(-k) \sum_{n=0}^{N-1} e^{2\pi i \frac{ukn}{N}}.$$

Because $(u, N) = 1$, the summation vanishes, unless $k = 0$, in which case it equals N, proving $\mathbf{x}_u$ satisfies ideal autocorrelation.

Theorem 6.3 *If $\mathbf{x}_u$ and $\mathbf{x}_v$ are unit discrete chirps in $\mathbb{C}^N$ such that*

$$(u - v, N) = 1,$$

then $(\mathbf{x}_u, \mathbf{x}_v)$ satisfies ideal correlation.

Proof By Theorem 6.1,

$$|(\mathbf{x}_u \circ \mathbf{x}_v)(k)|^2 = \sum_{n=0}^{N-1}\sum_{m=0}^{N-1} x_u(n)x_u^*(m)x_v^*(n-k)x_v(m-k)$$

$$= \sum_{n=0}^{N-1}\sum_{m=0}^{N-1} x_u(n)x_u^*(m)x_v^*(n)x_v(m)e^{2\pi i \frac{kv(n-m)}{N}}.$$

By the change of variables $r = n - m$ and Theorem 6.1,

$$|(\mathbf{x}_u \circ \mathbf{x}_v)(k)|^2 = \sum_{r=0}^{N-1} x_u^*(-r)x_v(-r)e^{2\pi i \frac{kvr}{N}} \sum_{n=0}^{N-1} e^{2\pi i \frac{(u-v)rn}{N}}.$$

Because $(u - v, N) = 1$, the second summation vanishes, unless $r = 0$, in which case it equals N, implying

$$|(\mathbf{x}_u \circ \mathbf{x}_v)(k)|^2 = N, \qquad 0 \le k < N,$$

and showing that $(\mathbf{x}_u, \mathbf{x}_v)$ satisfies ideal correlation.

Example 6.2 For $N = 225$, $\mathbf{x}_2$, $\mathbf{x}_4$ and $\mathbf{x}_7$ with discrete carrier frequencies $0, 0$ and $\frac{1}{2}$, respectively, satisfy ideal autocorrelation. $(\mathbf{x}_2, \mathbf{x}_4)$ satisfies ideal correlation while $(\mathbf{x}_2, \mathbf{x}_5)$ does not. Figure 6.4 displays the autocorrelations of one of $\mathbf{x}_2$, $\mathbf{x}_4$ and $\mathbf{x}_7$, which are identical. Figure 6.5 displays the acyclic autocorrelations, which can vary. Figure 6.6 displays the cross correlations and Figure 6.7 displays the acyclic cross correlations.

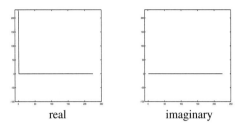

Fig. 6.4. Autocorrelation of a unit discrete chirp

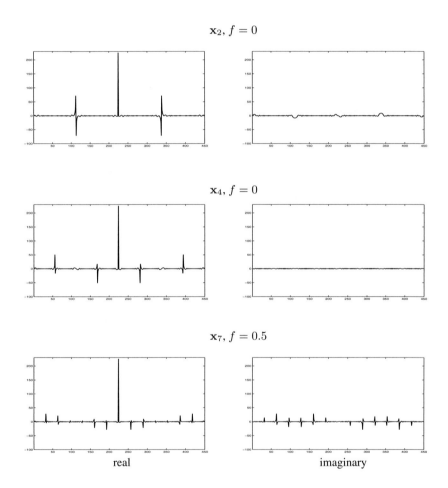

Fig. 6.5. Acyclic autocorrelations of unit discrete chirps

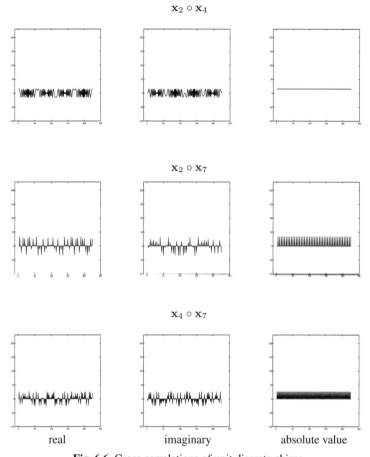

Fig. 6.6. Cross correlations of unit discrete chirps

An *echo* of a vector $\mathbf{x} \in \mathbb{C}^N$ is any linear combination,

$$\mathbf{e} = \sum_{r=0}^{N-1} \alpha(r) S_N^r \mathbf{x}.$$

In Chapter 14 we will define a more realistic model of an echo based on zero-padded vectors.

By Theorem 6.2, if $\mathbf{x}_u$ is a unit discrete chirp in $\mathbb{C}^N$, then the collection of shifts

$$\{S_N^r \mathbf{x}_u : 0 \leq r < N\}$$

is an orthogonal basis of $\mathbb{C}^N$. For $\mathbf{e} \in \mathbb{C}^N$, we have the expansion

$$\mathbf{e} = \sum_{r=0}^{N-1} \alpha(r) S_N^r \mathbf{x}_u,$$

$$\mathbf{x}_2 \circ \mathbf{x}_4 \qquad\qquad \mathbf{x}_2 \circ \mathbf{x}_7$$

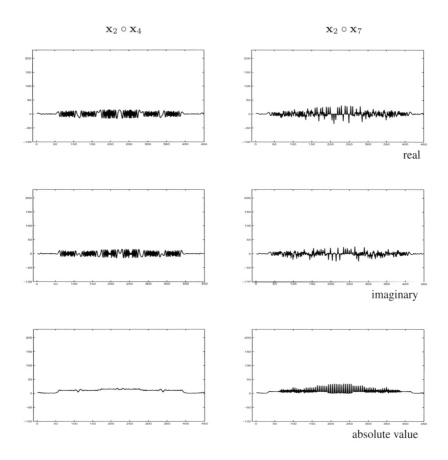

real

imaginary

absolute value

Fig. 6.7. Acyclic cross correlations of unit discrete chirps

where the coefficients of the expansion can be computed by

$$\alpha(r) = \frac{1}{N} \langle \mathbf{e}, S_N^r \mathbf{x}_u \rangle, \qquad 0 \leq r < N.$$

The *exponential vectors* $\mathbf{e}_m \in \mathbb{C}^N, 0 \leq m < N$, defined by

$$\mathbf{e}_m = [w^{mn}]_{0 \leq n < N}, \qquad w = e^{2\pi i \frac{1}{N}},$$

form an orthonormal basis of $\mathbb{C}^N$. Expansions of vectors in $\mathbb{C}^N$ over the exponential vectors are called *Fourier expansions*. In the following discussion we transform the expansion of $\mathbf{e}$ over the shifts of $\mathbf{x}_u$ into a Fourier expansion. The transform is called deramping or *dechirping*. The coefficients of the discrete chirp expansion and the dechirped Fourier expansion are related by a diagonal matrix multiplication.

For $0 \leq n < N$

$$e(n) = \sum_{r=0}^{N-1} \alpha(r) x_u(n - r).$$

By Theorem 6.1, because $(u, N) = 1$, the change of variables $m = -ur$ leads to the Fourier expansion

$$x_u(n)^* e(n) = \sum_{m=0}^{N-1} \alpha \left(-u^{-1}m\right) x_u \left(u^{-1}n\right) e^{2\pi i u \frac{nm}{N}},$$

where $-u^{-1}m$ and $u^{-1}n$ are taken modulo N.

Signal sets of discrete chirps having the same periodicity can be constructed by sampling chirp pulses having the same time duration and sampling rate, but varying chirp rates and carrier frequencies. Fix a time duration T and choose real numbers γ_j and ν_j, $1 \leq j < J$, such that $M_j = \gamma_j T^2$, $1 \leq j < J$, are positive integers having a common divisor $N > 1$ and

$$M_j + 2\nu_j T \in 2\mathbb{Z}, \qquad 1 \leq j < J.$$

Setting

$$u_j = \frac{M_j}{N} \text{ and } f_j = \nu_j T, \qquad 1 \leq j < J,$$

we can form the J discrete chirps of period N,

$$x_{u_j}(n) = e^{\pi i u_j \frac{n^2}{N}} e^{2\pi i f_j \frac{n}{N}}, \qquad 1 \leq j < J.$$

Example 6.3 Set $T = 40$. Table 6.1 lists the parameters for a set of 5 discrete chirps of period 40.

Table 6.1. Discrete chirp parameters of same period.

j	γ_j	M_j	u_j	ν_j
1	.500	800	20	.025
2	.250	400	10	.050
3	.200	320	8	401.025
4	.125	200	5	.525
5	.100	160	4	41.025

Because $M_j \in 2\mathbb{Z}$, $1 \leq j \leq 5$, to meet the condition

$$M_j + 2\nu_j T \in 2\mathbb{Z},$$

we must choose ν_j to be integer multiples of $\frac{1}{40}$. The periodic set of chirps are displayed in Figures 6.8–6.9.

$$\gamma_1 = .500, M_1 = 800, \nu_1 = .025$$

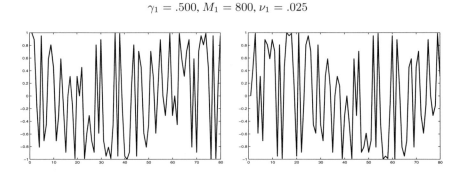

$$\gamma_2 = .250, M_2 = 400, \nu_2 = .050$$

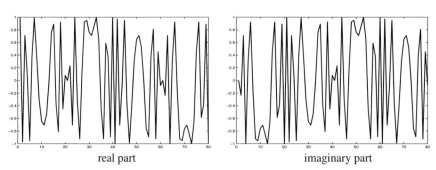

$$\gamma_3 = .200, M_3 = 320, \nu_3 = 401.025$$

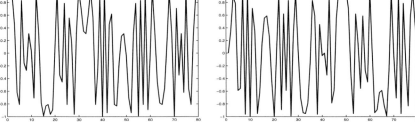

real part   imaginary part

Fig. 6.8. Periodic discrete chirp sets

6.2 Fourier Transform of Discrete Chirps

In this section we derive a formula for the N-point Fourier transform of the discrete chirp of period N,

$$x_u(n) = e^{\pi i u \frac{n^2}{N}} e^{2\pi i f \frac{n}{N}}, \qquad n \in \mathbb{Z}.$$

$$\gamma_4 = .125,\ M_3 = 200,\ \nu_3 = .525$$

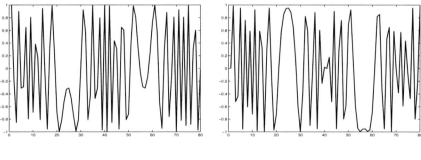

$$\gamma_5 = .100,\ M_4 = 160,\ \nu_4 = 41.025$$

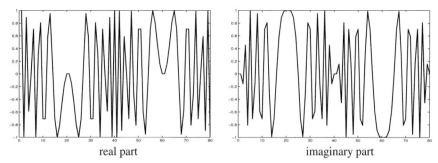

real part   imaginary part

Fig. 6.9. Periodic discrete chirp sets

A key element of the formula is the evaluation of certain constants in terms of quadratic
Gauss sums [7]. The use of generalized Gauss sums similar to the sums considered
here occur in [25] in connection with the computation of the Zak transform of the
time-continuous chirp $e^{\pi i \frac{p}{q} t^2}$ with p and q nonzero integers.

In the definition of x_u, u is a non-negative integer, but we do not assume, unless
otherwise specified, that x_u is a unit discrete chirp. Set

$$R = (u, N)$$

and

$$v = \frac{u}{R} \quad \text{and} \quad M = \frac{N}{R}.$$

We have $(v, M) = 1$. We show that, up to phase multiplication, x_u is periodic modulo
M and up to translation, the N-point Fourier transform of x_u vanishes except at
multiples of R. As a special case, the N-point Fourier transform of a unit discrete
chirp is, up to a constant multiple, a unit discrete chirp.

A sequence x satisfies the *chirp condition* of *chirp rate* u and *period* N if

$$x(0) = 1 \quad \text{and} \quad |x(n)| = 1,$$

$$x(n + N) = x(n),$$

and

$$x(n + m) = x(n)x(m)e^{2\pi i u \frac{nm}{N}}, \qquad n,\, m \in \mathbb{Z}.$$

By Theorem 6.1, x_u satisfies the chirp condition with chirp rate u and period N.

For the remainder of this section we assume x satisfies the chirp condition with chirp rate u and period N. Identify x with the vector

$$\mathbf{x}_N = [x(n)]_{0 \le n < N},$$

and denote by $\widehat{\mathbf{x}}_N$ the N-point Fourier transform of $\mathbf{x}_N$.

$$\widehat{\mathbf{x}}_N(r) = \sum_{n=0}^{N-1} x(n)e^{2\pi i r \frac{n}{N}}, \qquad 0 \le r < N.$$

Then

$$\widehat{\mathbf{x}}_N(r+s)\widehat{\mathbf{x}}_N^*(s) = \sum_{n=0}^{N-1} \sum_{m=0}^{N-1} x(n)x^*(m)e^{2\pi i \frac{rn}{N}} e^{2\pi i \frac{(n-m)s}{N}}.$$

By the change of variables, $l = n - m$,

$$\widehat{\mathbf{x}}_N(r+s)\widehat{\mathbf{x}}_N^*(s) = \sum_{l=0}^{N-1} x^*(-l)e^{2\pi i \frac{ls}{N}} \sum_{n=0}^{N-1} e^{2\pi i \frac{(ul+r)}{N}n}.$$

The summation over n vanishes unless

$$ul \equiv -r \mod N,$$

in which case it equals N, proving

$$\widehat{\mathbf{x}}_N(r+s)\widehat{\mathbf{x}}_N^*(s) = N \sum_{ul \equiv -r \bmod N} x^*(-l)e^{2\pi i \frac{s}{N}l}.$$

Theorem 6.4 *If $(u, N) = 1$, then*

$$|\widehat{\mathbf{x}}_N(s)|^2 = N,$$

and

$$\widehat{\mathbf{x}}_N(r) = \widehat{\mathbf{x}}_N(0)x^*\left(u^{-1}r\right), \qquad 0 \le s,\, r < N,$$

where $u^{-1}r$ is taken modulo N.

Proof Since $(u, N) = 1$,

$$l = -u^{-1}r \mod N,$$

is the unique solution of

$$ul \equiv -r \mod N,$$

and

$$\widehat{\mathbf{x}}_N(r+s)\widehat{\mathbf{x}}^*(s) = Nx^* \left(-u^{-1}r\right) e^{-2\pi i \frac{u^{-1}rs}{N}}.$$

Setting $r = 0$ proves the first formula and setting $s = 0$ proves the second, completing the proof.

Theorem 6.4 will be extended to the case

$$R = (u, N) \geq 1.$$

As

$$x(m+n) = x(m)x(n)e^{2\pi i v \frac{mn}{M}}, \qquad m, n \in \mathbb{Z},$$

if x is periodic modulo M, we can apply Theorem 6.4 with u replaced by v and N replaced by M. We have

$$x(n+M) = x(M)x(n), \qquad x \in \mathbb{Z}.$$

Continuing,

$$x(n+rM) = x(M)^r x(n), \qquad n, r \in \mathbb{Z}.$$

As x is periodic modulo N,

$$x(M)^R = 1.$$

As a result x satisfies the chirp condition with chirp rate v and period M if and only if

$$x(M) = 1.$$

Theorem 6.4 leads to the next result.

Theorem 6.5 *If $x(M) = 1$, then*

$$|\widehat{\mathbf{x}}_M(s)|^2 = M, \qquad 0 \leq s < M,$$

and

$$\widehat{\mathbf{x}}_M(r) = \widehat{\mathbf{x}}_M(0)x^* \left(v^{-1}r\right), \qquad 0 \leq r < M,$$

where $v^{-1}r$ is taken modulo M.

We require the following general result from the theory of the Fourier transform of periodic data [40]. Suppose y is a sequence of period M and N is a multiple of M, $N = RM$. Then

$$\widehat{\mathbf{y}}_N(kR) = R\widehat{\mathbf{y}}_M(k), \qquad 0 \leq k < M,$$

and the remaining components of $\widehat{\mathbf{y}}_N$ vanish. Combining this result with Theorem 6.5 we have the next result.

Theorem 6.6 *If $x(M) = 1$, then*

$$\widehat{\mathbf{x}}_N(kR) = R\widehat{\mathbf{x}}_M(0)x^* \left(v^{-1}k\right), \qquad 0 \leq k < M,$$

where $v^{-1}k$ is taken modulo M and the remaining components of $\widehat{\mathbf{x}}_N$ vanish.

We can apply Theorem 6.6 to the computation of the N-point Fourier transform $\widehat{\mathbf{x}}_u$ of

$$\mathbf{x}_u(n) = [x_u(n)]_{0 \leq n < N} \,.$$

There are three cases.

Case 1 uN is even and vM is even. Since $\mathbf{x}_u$ is periodic modulo N, we have $f \in \mathbb{Z}$. Set

$$y(n) = e^{\pi i v \frac{n^2}{M}}, \quad n \in \mathbb{Z}.$$

Because

$$x_u(n) = y(n)e^{2\pi i f \frac{n}{N}}, \quad f \in \mathbb{Z},$$

we have

$$\widehat{\mathbf{x}}_u(r - f) = \widehat{\mathbf{y}}_N(r), \quad 0 \leq r < N.$$

y satisfies the chirp condition with chirp rate v and period M. By Theorem 6.6,

$$\widehat{\mathbf{y}}_N(kR) = R\widehat{\mathbf{y}}_M(0)e^{\pi i(-v^{-1})\frac{k^2}{M}}, \quad 0 \leq k < M.$$

Combining results,

$$\widehat{\mathbf{x}}_u(kR - f) = R\widehat{\mathbf{y}}_M(0)e^{\pi i(-v^{-1})\frac{k^2}{M}}, \quad 0 \leq k < M,$$

and the remaining components of $\widehat{\mathbf{x}}_u$ vanish.

Case 2 uN is even and vM is odd. N must be even. Set

$$f = \frac{N}{2} + f_0, \quad f_0 \in \mathbb{Z},$$

and

$$z(n) = e^{\pi i v \frac{n^2}{M}}(-1)^n, \quad n \in \mathbb{Z}.$$

Because

$$x_u(n) = z(n)e^{2\pi i f_0 \frac{n}{N}}, \quad f_0 \in \mathbb{Z},$$

we have

$$\widehat{\mathbf{x}}_u(r - f_0) = \widehat{\mathbf{z}}_N(r), \quad 0 \leq r < N.$$

z satisfies the chirp condition with chirp rate v and period M. Arguing as above,

$$\widehat{\mathbf{x}}_u(kR - f_0) = R\widehat{\mathbf{z}}_M(0)e^{\pi i(-v^{-1})\frac{k^2}{M}}(-1)^{v^{-1}k}, \quad 0 \leq k < M,$$

and the remaining components of $\widehat{\mathbf{x}}_u$ vanish.

Case 3 uN is odd. Because $2f$ is odd and N is odd, we can write

$$f = \frac{N}{2} + f_1, \quad f_1 \in \mathbb{Z}.$$

Set

$$w(n) = e^{\pi i v \frac{n^2}{M}}(-1)^n.$$

Because

$$x_u(n) = w(n)e^{2\pi i f_1 \frac{n}{N}}, \qquad f_1 \in \mathbb{Z},$$

we have

$$\widehat{\mathbf{x}}_u(r - f_1) = \widehat{\mathbf{w}}_N(r), \qquad 0 \le r < N.$$

w satisfies the chirp condition with chirp rate v and period M. Arguing as above,

$$\widehat{\mathbf{x}}_u(kR - f_1) = R\widehat{\mathbf{w}}_M(0)e^{\pi i\left(-v^{-1}\right)\frac{k^2}{M}}(-1)^{v^{-1}k}, \qquad 0 \le k < M,$$

and the remaining components of $\widehat{\mathbf{x}}_u$ vanish.

We summarize the three cases in Table 6.3. $\widehat{\mathbf{x}}_u$ vanishes at the components not indicated. The special case $(u, N) = 1$ is summarized in Table 6.2.

Table 6.2. Fourier transform of discrete chirps, $R = (u, N) = 1$

uN	f	$\widehat{\mathbf{x}}_u, 0 \le k < N$
even		$c_0 e^{\pi i\left(-u^{-1}\right)\frac{k^2}{N}} e^{2\pi i\left(-u^{-1}\right)\frac{fk}{N}}$
odd	$f = \frac{N}{2} + f_0$	$c_1 e^{\pi i\left(-u^{-1}\right)\frac{k^2}{N}} e^{2\pi i\left(-u^{-1}\right)\frac{f_0 k}{N}}$

Table 6.3. Fourier transform of discrete chirps

uN	vM	f	$\widehat{\mathbf{x}}_u(kR - f), 0 \le k < M$
even	even		$R\widehat{\mathbf{y}}_M(0)e^{\pi i\left(-v^{-1}\right)\frac{k^2}{M}}$
even	odd	$f = \frac{N}{2} + f_0$	$R\widehat{\mathbf{z}}_M(0)e^{\pi i\left(-v^{-1}\right)\frac{k^2}{M}}(-1)^{v^{-1}k}$
odd		$f = \frac{N}{2} + f_1$	$R\widehat{\mathbf{w}}_M(0)e^{\pi i\left(-v^{-1}\right)\frac{k^2}{M}}(-1)^{v^{-1}k}$

$$c_0 = e^{\pi i\left(-u^{-1}\right)\frac{f^2}{N}}\widehat{\mathbf{y}}_N(0)$$

and

$$c_1 = e^{\pi i\left(-u^{-1}\right)\frac{f_0^2}{N}}(-1)^{u^{-1}(k+f_0)}\widehat{\mathbf{z}}_N(0),$$

where u^{-1} is taken modulo N. By Table 6.1 the N-point Fourier transform of a unit discrete chirp is, up to a constant multiple, a unit discrete chirp.

6.3 Gauss Sums

For $v, M \in \mathbb{Z}$, $(v, M) = 1$, the *quadratic Gauss sum* $G(v, M)$ is defined by

$$G(v, M) = \sum_{n=0}^{M-1} e^{2\pi i v \frac{n^2}{M}}.$$

In the following discussion we relate the constants $\widehat{\mathbf{y}}_M(0)$, $\widehat{\mathbf{z}}_M(0)$ and $\widehat{\mathbf{w}}_M(0)$ to quadratic Gauss sums. Consider first

$$\widehat{\mathbf{y}}_M(0) = \sum_{n=0}^{M-1} e^{\pi i v \frac{n^2}{M}},$$

where $(v, M) = 1$ and vM is even. We have two cases.

Case 1 v is even and M is odd.

$$\widehat{\mathbf{y}}_M(0) = \sum_{n=0}^{M-1} e^{2\pi i \frac{v}{2} \frac{n^2}{M}} = G\left(\frac{v}{2}, M\right).$$

Case 2 v is odd and M is even.

$$\widehat{\mathbf{y}}_M(0) = \sum_{n=0}^{M-1} e^{2\pi i v \frac{n^2}{2M}}.$$

Because

$$G(v, 2M) = \sum_{n=0}^{2M-1} e^{2\pi i v \frac{n^2}{2M}} = 2 \sum_{n=0}^{M-1} e^{2\pi i v \frac{n^2}{2M}},$$

we have

$$\widehat{\mathbf{y}}_M(0) = \frac{1}{2} G(v, 2M).$$

Consider now

$$\widehat{\mathbf{z}}_M(0) = \sum_{n=0}^{M-1} e^{\pi i v \frac{n^2}{M}} (-1)^n = \sum_{n=0}^{M-1} e^{2\pi i v \frac{n^2}{2M}} (-1)^n,$$

with vM odd. We can write

$$\widehat{\mathbf{z}}_M(0) = \frac{1}{2} \sum_{n=0}^{2M-1} e^{2\pi i v \frac{n^2}{2M}} (-1)^n.$$

Summing over even and odd n separately,

$$2\widehat{\mathbf{z}}_M(0) = \sum_{n=0}^{M-1} e^{2\pi i 2v \frac{n^2}{M}} - \sum_{n=0}^{M-1} e^{2\pi i v \frac{(2n+1)^2}{2M}}.$$

Because

$$\sum_{n=0}^{2M-1} e^{2\pi i v \frac{n^2}{2M}} = 0,$$

we have

$$\sum_{n=0}^{M-1} e^{2\pi i v \frac{(2n+1)^2}{2M}} = -\sum_{n=0}^{M-1} e^{2\pi i \frac{v(2n)^2}{2M}} = -\sum_{n=0}^{M-1} e^{2\pi i 2v \frac{n^2}{M}}.$$

As a result,

$$2\widehat{\mathbf{z}}_M(0) = 2\sum_{n=0}^{M-1} e^{2\pi i 2v \frac{n^2}{M}} = 2G(2v, M)$$

and

$$\widehat{\mathbf{z}}_M(0) = G(2v, M).$$

$\widehat{\mathbf{w}}_M(0)$ has the same form as $\widehat{\mathbf{z}}_M(0)$ and vM is odd, implying

$$\widehat{\mathbf{w}}_M(0) \quad G(2v, M).$$

These results are organized in Table 6.4.

Table 6.4. Gauss sums

v	M	Gauss sum
even	odd	$\widehat{\mathbf{y}}_M(0) = G\left(\frac{v}{2}, M\right)$
odd	even	$\widehat{\mathbf{y}}_M(0) = \frac{1}{2}G(v, 2M)$
odd	odd	$\widehat{\mathbf{z}}_M(0) = G(2v, M)$
odd	odd	$\widehat{\mathbf{w}}_M(0) = G(2v, M)$

6.4 Computation of Gauss Sums

The final step in the evaluation of the N-point Fourier transform of the discrete chirp $\mathbf{x}_u$ of period N is the evaluation of the Gauss sums [4]. For this purpose we must first define the Legendre symbol and the Jacobi symbol.

Suppose p is an odd prime and $a \in U_p$. a is called a *quadratic residue* mod p if the congruence

$$x^2 \equiv a \mod p$$

has a solution. Otherwise a is called a *quadratic nonresidue* mod p. Define the *Legendre symbol* $\left(\frac{a}{p}\right)$ by

$$\left(\frac{a}{p}\right) = \begin{cases} 1, & a \text{ is a quadratic residue mod } p, \\ -1, & a \text{ is a quadratic nonresidue mod } p. \end{cases}$$

The definition of Legendre symbol is usually extended to all integers by setting

$$\left(\frac{a}{p}\right) = 0, \quad p \text{ divides } a.$$

Suppose $M > 0$ is an odd integer and

$$M = p_1^{r_1} \cdots p_k^{r_k}$$

is its prime factorization. For any $v \in \mathbb{Z}$, the *Jacobi symbol* $\left(\frac{v}{M}\right)$ is defined by

$$\left(\frac{v}{M}\right) = \left(\frac{v}{p_1}\right)^{r_1} \cdots \left(\frac{v}{p_k}\right)^{r_k},$$

where the symbols on the right are all Legendre symbols.

Gauss sums are computed by the following formulas:

$$G(v, M) = \begin{cases} \left(\frac{M}{v}\right)(1 + i^v)\sqrt{M}, & M \equiv 0 \bmod 4, \\ \left(\frac{v}{M}\right)\sqrt{M}, & M \equiv 1 \bmod 4, \\ 0, & M \equiv 2 \bmod 4 \\ \left(\frac{v}{M}\right)i\sqrt{M}, & M \equiv 3 \bmod 4, \end{cases}$$

where $\left(\frac{M}{v}\right)$ and $\left(\frac{v}{M}\right)$ are Jacobi symbols.

The evaluation of Gauss sums is one of the most significant events of 19th century mathematics. Gauss, in 1811 in his famous paper [19], derived formulas computing classical Gauss sums and used these formulas to give a simple proof of his law of quadratic reciprocity. Since then, Gauss along with some of the most famous names in mathematics have extended these results to Gauss sums of multiplicative characters and higher reciprocity laws. These results significantly impact number theory, theta function theory, digital signal processing and coding theory. The derivations of these generalizations are beyond the scope of this text, but some of these results are potentially useful in discrete chirp analysis and signal design.

7

Zak Transform

For over fifty years, beginning with the fundamental work of D. Gabor [17], time-frequency representations have provided deep insights into a wide range of problems in signal analysis, processing and design. This effort deals with the digital aspect of these representations. The underlying premise is that the discrete Zak transform is the primary time-frequency representation. Many of the best algorithms for computing the ambiguity function, the Wigner distribution and Gabor expansions pass through the Zak transform.

The finite Zak transform of a vector determines a two-dimensional time-frequency representation of the vector called the *Zak space representation*. It is a natural setting for studying discrete linear frequency modulated chirps as they are intrinsically time-frequency signals which have compact representations in Zak space. The Zak transform is named after J. Zak, who studied it systematically [43] for applications in solid state physics. Independently, A. Weil described a similar transform, later called the Weil–Brezin transform, in his work [42] on theta function theory and representations of the Heisenberg group. Its implicit applications to analytic number theory can be found in earlier works of C.L. Siegel [34]. It is intimately linked to the Poisson summation formula, which describes the relationship between periodic and decimated data under the Fourier transform.

Its role in digital signal processing begins in earnest with the work of A.J.E.M. Janssen [25], but it previously served as an essential step in the Cooley–Tukey fast Fourier transform (CT FFT) algorithm [13, 40]. Since Janssen's work, the Zak transform has been used in many applications in signal and echo analysis, ambiguity function and Gabor expansion analysis and computation [16, 14, 30, 38].

In several recent works [8, 9, 37, 39], Zak space representations have provided an image for processing and analyzing chirp reflections from multiple targets and dielectric materials. This last application motivates much of the signal design strategy in this work.

In this chapter we define the finite Zak transform and discuss some of its basic properties. The finite Zak transform provides an image of a vector and operations on the image. We study two of the most important operations, shift and the finite Fourier transform. The results on shift will be used in Chapter 8 to derive the Zak

M. An et al., *Ideal Sequence Design in Time-Frequency Space*,
DOI 10.1007/978-0-8176-4738-4_7,
© Birkhäuser Boston, a part of Springer Science+Business Media, LLC 2009

space (ZS) correlation formula describing the finite Zak transform of the correlation of two vectors. The ZS realization of the finite Fourier transform of a vector is a geometric operation on the ZS representation of this vector. This result has been used in [37, 39] to construct purely in Zak space an orthogonal basis diagonalizing the finite Fourier transform.

Throughout this chapter, $N = LK$, where L and K are positive integers. Unless otherwise specified, $F = F(L)$, $D = D_L$, $S = S_L$, $R = R_L$ and $\mathbf{y} \in \mathbb{C}^N$. We recall several definitions from Chapter 4.

Identify $\mathbb{C}^L \times \mathbb{C}^K$ with the space of $L \times K$ complex matrices. For $\mathbf{y} \in \mathbb{C}^N$ consider the vectors $\mathbf{y}_k \in \mathbb{C}^L, 0 \le k < K$,

$$\mathbf{y}_k = [y_{k+lK}]_{0 \le l < L}, \qquad 0 \le k < K,$$

and the $L \times K$ matrix $M\mathbf{y}$,

$$M\mathbf{y} = [\mathbf{y}_0 \quad \cdots \quad \mathbf{y}_{K-1}].$$

Define

$$Z\mathbf{y} = FM\mathbf{y}.$$

$Z\mathbf{y}$ is called the $L \times K$ ZS *representation* of $\mathbf{y}$. The mapping

$$Z : \mathbb{C}^N \longrightarrow \mathbb{C}^L \times \mathbb{C}^K$$

is called the $L \times K$ *finite Zak transform*.

When the dependence on L or $L \times K$ must be distinguished, we write M_L or $M_{L \times K}$ for M and Z_L or $Z_{L \times K}$ for Z. $Z\mathbf{y}$ can be computed by first computing

$$(I_K \otimes F) P(N, K)\mathbf{y}$$

and then placing contiguous blocks of length L vectors in an $L \times K$ array.

We recover $M\mathbf{y}$ from $Z\mathbf{y}$ by

$$M\mathbf{y} = F^{-1}Z\mathbf{y}.$$

Placing the usual inner product on a complex $L \times K$ array, we have

$$||Z\mathbf{y}||^2 = L||\mathbf{y}||^2.$$

Up to scalar factor L, Z is a linear isometry from $\mathbb{C}^N$ onto $\mathbb{C}^L \times \mathbb{C}^K$.

The information contained in $\mathbf{y} \in \mathbb{C}^N$ is the same as that contained in $Z\mathbf{y}$, but is presented differently. $Z\mathbf{y}$ is a two-dimensional image of $\mathbf{y}$. This image changes as L changes. $Z_N\mathbf{y}$ is $F(N)\mathbf{y}$, while $Z_1\mathbf{y} = \mathbf{y}^T$. For a specific $\mathbf{y} \in \mathbb{C}^N$, we try to choose L such that $Z_L\mathbf{y}$ is sparse. This will be the case for discrete chirps in Chapter 9 and is one of the principles of sequence design in this text.

Example 7.1

$$Z\mathbf{e}_0^N = [\mathbf{1} \quad \mathbf{0} \quad \cdots \quad \mathbf{0}], \qquad \mathbf{1} = 1^L, \quad \mathbf{0} = 0^L.$$

7.1 ZS Representation of Shifts

Set

$$Z\mathbf{y} = [Y_0 \ Y_1 \ \cdots \ Y_{K-1}], \qquad Y_k \in \mathbb{C}^L.$$

The shift matrix S_N is the basic building block for the most frequently occurring time-domain operations in digital signal processing, convolution and correlation. By the formula

$$D_N = F(N)S_N F(N)^{-1},$$

the phase matrix D_N plays an equally important role in frequency-domain operations. In the following two theorems we derive formulas for the ZS realizations of the actions of S_N and D_N on $\mathbb{C}^N$. These formulas will be used to derive the ZS correlation formula in Chapter 8 and the ZS realization of the N-point Fourier transform in Section 7.2.

By Theorem 3.1,

$$M\left(S_N^k \mathbf{y}\right) = [S\mathbf{y}_{K-k} \ \cdots \ S\mathbf{y}_{K-1} \ \mathbf{y}_0 \ \cdots \ \mathbf{y}_{K-k-1}]$$

and

$$M\left(S_N^{k+lK}\mathbf{y}\right) = S^l M\left(S_N^k \mathbf{y}\right), \qquad 0 \le k < K, \ 0 \le l < L.$$

Because $FSF^{-1} = D$, we have

$$Z\left(S_N^k \mathbf{y}\right) = FM\left(S_N^k \mathbf{y}\right) = [DY_{K-k} \ \cdots \ DY_{K-1} \ Y_0 \ \cdots Y_{K-k-1}]$$

and

$$Z\left(S_N^{k+lK}\mathbf{y}\right) = D^l Z\left(S_N^k \mathbf{y}\right), \qquad 0 \le k < K, \ 0 \le l < L,$$

proving the next result.

Theorem 7.1

$$Z\left(S_N^k \mathbf{y}\right) = [DY_{K-k} \ \cdots \ DY_{K-1} \ Y_0 \ \cdots Y_{K-k-1}]$$

and

$$Z\left(S_N^{k+lK}\mathbf{y}\right) = D^l Z\left(S_N^k \mathbf{y}\right), \qquad 0 \le k < K, \ 0 \le l < L.$$

Set

$$\mathbf{e}_n = \mathbf{e}_n^N = S_N^n \mathbf{e}_0, \qquad 0 \le n < N.$$

Writing $F = F(L)$ in terms of its column vectors

$$F = [F_0 \ \cdots F_{L-1}],$$

we have

$$F_l = D^l \mathbf{1}, \qquad 0 \le l < L.$$

Corollary 7.1 *For $0 \le k < K$ and $0 \le l < L$*

$$Z\mathbf{e}_k = \begin{bmatrix} \mathbf{0} & \cdots & \mathbf{1} & \cdots & \mathbf{0} \end{bmatrix},$$

and

$$Z\mathbf{e}_{k+lK} = \begin{bmatrix} \mathbf{0} & \cdots & F_l & \cdots \mathbf{0} \end{bmatrix}, \qquad \mathbf{0} = \mathbf{0}^L, \ \mathbf{1} = \mathbf{1}^L,$$

where $\mathbf{1}$ and F_l are in the k-th column.

Example 7.2 $L = 3$ and $K = 4$.

$$Z\mathbf{e}_0 = \begin{bmatrix} \mathbf{1} & \mathbf{0} & \mathbf{0} & \mathbf{0} \end{bmatrix},$$

$$Z\mathbf{e}_1 = \begin{bmatrix} \mathbf{0} & \mathbf{1} & \mathbf{0} & \mathbf{0} \end{bmatrix},$$

$$Z\mathbf{e}_2 = \begin{bmatrix} \mathbf{0} & \mathbf{0} & \mathbf{1} & \mathbf{0} \end{bmatrix},$$

$$Z\mathbf{e}_3 = \begin{bmatrix} \mathbf{0} & \mathbf{0} & \mathbf{0} & \mathbf{1} \end{bmatrix},$$

and

$$Z\mathbf{e}_4 = D\begin{bmatrix} \mathbf{1} & \mathbf{0} & \mathbf{0} & \mathbf{0} \end{bmatrix}.$$

Example 7.3 For $N = 12 \times 10$, $Z_{12}\mathbf{e}_k$ and the Zak transform of shifts are plotted in Figure 7.1.

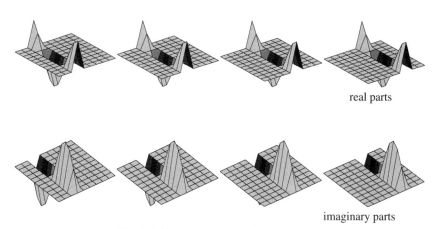

real parts

imaginary parts

Fig. 7.1. $Z_{12}\mathbf{e}_{13}$ and the Zak transform of shifts

Since $Z_{12}\mathbf{e}_{13}$ and its 3 shifts have disjoint supports, there is no arithmetic in summing them, it is merely a juxtaposition. This is displayed in Figure 7.2.

$Z_{12}\mathbf{e}_{1+lK}$ is supported only on the first column, $0 \le l < 10$. The values on the first column of $Z_{12}\mathbf{e}_{1+lK}$, $0 \le l \le 3$, are plotted in Figure 7.3.

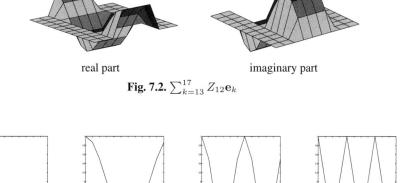

real part   imaginary part

Fig. 7.2. $\sum_{k=13}^{17} Z_{12}\mathbf{e}_k$

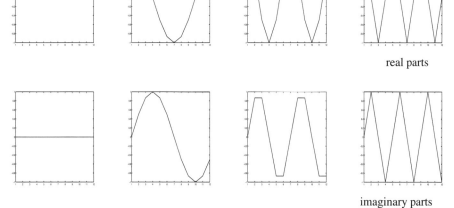

real parts

imaginary parts

Fig. 7.3. Nonzero values of $Z_{12}\mathbf{e}_1$ and the Zak transform of shifts by 10

Set $w = e^{2\pi i \frac{1}{N}}$ and

$$D_K(N) = D\left(\left[w^k\right]_{0 \le k < K}\right).$$

For $\mathbf{y} \in \mathbb{C}^N$, the k-th column of $M(D_N\mathbf{y})$ is

$$w^k D\mathbf{y}_k,$$

where $\mathbf{y}_k, 0 \le k < K$, are the K-decimated components of $\mathbf{y}$. Applying F, the k-th column of $Z(D_N\mathbf{y})$ is

$$F\left(w^k D\mathbf{y}_k\right) = w^k S^{-1}Y_k, \qquad 0 \le k < K,$$

and

$$Z(D_N\mathbf{y}) = S^{-1}\left[Y_0 \ \ wY_1 \ \ \cdots \ \ w^{K-1}Y_{K-1}\right] = S^{-1}(Z\mathbf{y})D_K(N).$$

Continuing,

$$Z\left(D_N^l \mathbf{y}\right) = S^{-l}(Z\mathbf{y})D_K(N)^l, \qquad 0 \le l < L,$$

and

$$Z\left(D_N^L \mathbf{y}\right) = Z\mathbf{y}D_K,$$

proving the following result.

Theorem 7.2

$$Z\left(D_N^l \mathbf{y}\right) = S^{-l}(Z\mathbf{y})D_K(N)^l,$$

and

$$Z\left(D_N^{l+kL}\mathbf{y}\right) = S^{-l}(Z\mathbf{y})D_K^k D_K(N)^l, \qquad 0 \le l < L, \ 0 \le k < K.$$

7.2 ZS Representation of Fourier Transform

We will now derive a formula for the $K \times L$ ZS representation of the finite Fourier transform $F(N)\mathbf{y}$ in terms of the $L \times K$ ZS representation of $\mathbf{y}$. The eventual goal is to realize the Fourier transform as a geometric action on Zak space. Set $\mathbf{0} = \mathbf{0}^L$, $\mathbf{1} = \mathbf{1}^L$, $E_k^K = \mathbf{e}_k^K, 0 \le k < K$, and

$$\mathbf{e}_n = \mathbf{e}_n^N, \qquad 0 \le n < N.$$

The n-th column of $F(N)$ is

$$F(N)\mathbf{e}_n = D_N^n \mathbf{1}^N, \qquad 0 \le n < N.$$

Theorem 7.3

$$Z_K\left(F(N)\mathbf{e}_n\right) = K\left[E_{K-k}^K \quad \cdots \quad E_{K-k}^K\right] D_L(N)^n,$$

where $n = k + lK, 0 \le k < K, 0 \le l < L$.

Proof By Theorem 7.2 with L and K interchanged,

$$Z_K\left(F(N)\mathbf{e}_n\right) = S_K^{-k} Z_K\left(\mathbf{1}^N\right) D_L(N)^n = K S_K^{-k}\left[E_0^K \quad \cdots \quad E_0^K\right] D_L(N)^n,$$

proving the theorem.

By Theorem 7.3 the $K \times L$ ZS representation of $F(N)\mathbf{e}_n$ is supported on the horizontal line,

$$\left[E_{K-k}^K \quad \cdots \quad E_{K-k}^K\right], \qquad n = k + lK.$$

In particular, $F(N)\mathbf{e}_n$ and $F(N)\mathbf{e}_{n'}$, with $n \equiv n' \bmod K$, have the same $K \times L$ ZS representation support. They are distinguished by their values on the horizontal line.

If $n \not\equiv n' \bmod K$, then the $K \times L$ ZS representations of $F(N)\mathbf{e}_n$ and $F(N)\mathbf{e}_{n'}$ have disjoint supports.

By Corollary 7.1,

$$R_K \left(Z_L \mathbf{e}_k\right)^T = S_K^{-k} \left[E_0^K \quad \cdots \quad E_0^K\right], \qquad 0 \le k < K.$$

Placing this result into the formula for $Z_K \left(F(N)\mathbf{e}_n\right)$, we have

$$Z_K \left(F(N)\mathbf{e}_n\right) = K R_K \left(Z_L \mathbf{e}_k\right)^T D_L(N)^n,$$

proving the following result.

Corollary 7.2

$$Z_K \left(F(N)\mathbf{e}_n\right) = K R_K \left(Z_L \mathbf{e}_k\right)^T D_L(N)^n,$$

where $n = k + lK, 0 \le k < K, 0 \le l < L.$

Example 7.4 $L = 3 = K, w = e^{2\pi i \frac{1}{9}}.$

$$Z_3 \left(F(9)\mathbf{e}_0\right) = 3 \begin{bmatrix} 1 & 1 & 1 \\ 0 & 0 & 0 \\ 0 & 0 & 0 \end{bmatrix},$$

$$Z_3 \left(F(9)\mathbf{e}_1\right) = 3 \begin{bmatrix} 0 & 0 & 0 \\ 0 & 0 & 0 \\ 1 & w & w^2 \end{bmatrix},$$

and

$$Z_3 \left(F(9)\mathbf{e}_2\right) = 3 \begin{bmatrix} 0 & 0 & 0 \\ 1 & w^2 & w^4 \\ 0 & 0 & 0 \end{bmatrix}.$$

Example 7.5 For $L = 8$ and $K = 10$, Figure 7.4 displays $Z_{10}F(80)\mathbf{e}_{32}$ and $Z_8\mathbf{e}_{13}$ as image plots. Time-reversal and transpose are easy to visualize. $D_{10}(80)^{32}$ is plotted in Figure 7.5.

Take $\mathbf{y} \in \mathbb{C}^N$ and write

$$\mathbf{y} = \sum_{n=0}^{N-1} y_n \mathbf{e}_n.$$

Set

$$Z_L \mathbf{y} = \left[Y_0 \quad \cdots \quad Y_{K-1}\right].$$

By linearity and Corollary 7.2,

$$Z_K \left(F(N)\mathbf{y}\right) = K R_K \sum_{n=0}^{N-1} y_n \left(Z_L \mathbf{e}_k\right)^T D_L(N)^n, \qquad k \equiv n \bmod K.$$

$Z_{10}F(80)\mathbf{e}_{32}$ $Z_8\mathbf{e}_{13}$

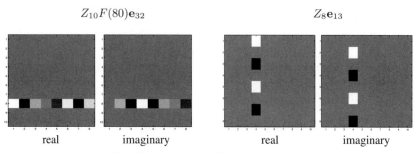

real   imaginary   real   imaginary

Fig. 7.4. Fourier transform and Zak transform

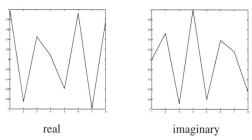

real   imaginary

Fig. 7.5. $D_{10}(80)^{32}$

Because

$$D_L(N)^n = D_L^l D_L(N)^k,$$

we have

$$Z_K(F(N)\mathbf{y}) = KR_K \sum_{k=0}^{K-1} (Z_L\mathbf{e}_k)^T \left(\sum_{l=0}^{L-1} y_{k+lK} D_L^l\right) D_L(N)^k$$

$$= KR_K \sum_{k=0}^{K-1} (Z_L\mathbf{e}_k)^T D(Y_k) D_L(N)^k,$$

where $D(Y_k)$ is the $L \times L$ diagonal matrix whose diagonal entries are given by Y_k, $0 \le k < L$. Then

$$Z_K(F(N)\mathbf{y}) = KR_K \sum_{k=0}^{K-1} \begin{bmatrix} E_k^K & \cdots & E_k^K \end{bmatrix} D(Y_k) D_L(N)^k.$$

The k-th row of

$$\sum_{k=0}^{K-1} \begin{bmatrix} E_k^K & \cdots & E_k^K \end{bmatrix} D(Y_k) D_L(N)^k$$

is

$$Y_k^T D_L(N)^k, \qquad 0 \leq k < K,$$

proving the next result.

Theorem 7.4

$$Z_K\left(F(N)\mathbf{y}\right) = K R_K \begin{bmatrix} Y_0^T \\ Y_1^T D_L(N) \\ \vdots \\ Y_{K-1}^T D_L(N)^{K-1} \end{bmatrix}.$$

The formula in Theorem 7.4 expresses the $K \times L$ ZS representation of $F(N)\mathbf{y}$ in terms of the $L \times K$ ZS representation of $\mathbf{y}$. In Section 7.3 we will use the ZS interchange theorem to write the $L \times K$ ZS representation of $F(N)\mathbf{y}$ in terms of the $L \times K$ ZS representation of $\mathbf{y}$.

Example 7.6 Define $\mathbf{x}$ by

$$x(t) = e^{-at^2/2}, \qquad a = \sqrt{\frac{2\pi}{256}}, \qquad -127 \leq t \leq 127.$$

Periodizing modulo 256 we have the shifted Gaussian

$$x_s(t) = e^{-at^2/2}, \qquad a = \sqrt{\frac{2\pi}{256}}, \qquad t \in \mathbb{Z}/256.$$

Figure 7.6 plots x, x_s and $F(256)\mathbf{x}_s$. Figures 7.7–7.8 display the Zak transforms of $\mathbf{x}_s$ and $F(256)\mathbf{x}_s$ as surf plots. The z-axes for the plots are enlarged to show more details. $Z_{16}\mathbf{x}_s$ vanishes only at $(8, 8)$ with a steep gradient and has no other values near zero. This property is important in several time-frequency expansions, including the Gabor expansion [14, 38].

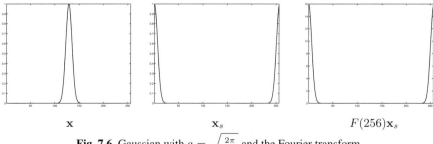

$$\mathbf{x} \qquad\qquad\qquad \mathbf{x}_s \qquad\qquad\qquad F(256)\mathbf{x}_s$$

Fig. 7.6. Gaussian with $a = \sqrt{\frac{2\pi}{256}}$ and the Fourier transform

$$Z_8 \mathbf{x}_s$$

$$\tfrac{1}{32} Z_{32} F(256) \mathbf{x}_s$$

$$Z_{32} \mathbf{x}_s$$

$$\tfrac{1}{8} Z_8 F(256) \mathbf{x}_s$$

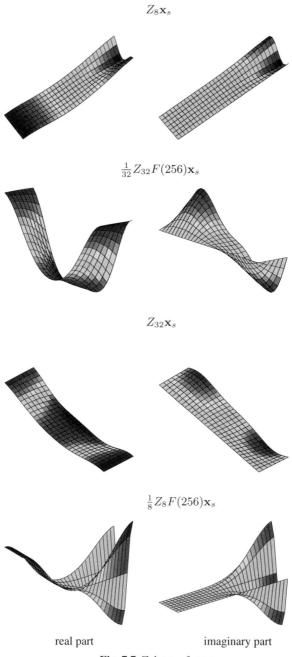

real part   imaginary part

Fig. 7.7. Zak transforms

$$Z_{16}\mathbf{x}_s$$

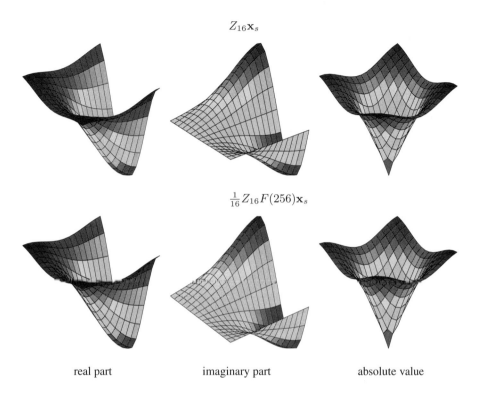

$$\tfrac{1}{16}Z_{16}F(256)\mathbf{x}_s$$

real part   imaginary part   absolute value

Fig. 7.8. Zak transforms

The formula in Theorem 7.4, up to scalar multiplication by K, computes $Z_K(F(N)\mathbf{y})$ by first transposing $Z_L\mathbf{y}$, then by twiddling the coefficients of $(Z_L\mathbf{y})^T$ by powers of $w = e^{2\pi i \frac{1}{N}}$, and then by time-reversing the rows of the twiddled $(Z_L\mathbf{y})^T$. Set

$$Y_k = [Y_{l,k}]_{0 \le l < L}, \qquad 0 \le k < K.$$

Then $Z_K(F(N)\mathbf{y})$ is

$$KR_K \begin{bmatrix} Y_{0,0} & Y_{1,0} & \cdots & Y_{L-1,0} \\ Y_{0,1} & Y_{1,1}w & \cdots & Y_{L-1,1}w^{L-1} \\ \vdots & & & \\ Y_{0,K-1} & Y_{1,K-1}w^{K-1} & \cdots & Y_{L-1,K-1}w^{(L-1)(K-1)} \end{bmatrix}.$$

Set

$$Z_K(F(N)\mathbf{y}) = [V_0 \ \cdots \ V_{L-1}]$$

and write

$$V_k = [V_{k,l}]_{0 \le l < L}, \qquad k \le l < K.$$

Corollary 7.3

$$V_{0,l} = KY_{l,0}, \qquad 0 \le l < L,$$

and

$$V_{k,l} = KY_{l,K-k}w^{l(K-k)}, \qquad 1 \le k < K, \, 0 \le l < L.$$

Example 7.7 $L = K = 2$. Suppose

$$\mathbf{y}^T = \begin{bmatrix} \frac{1}{2} & 1 & -\frac{1}{2} & 0 \end{bmatrix}.$$

Then

$$(F(4)\mathbf{y})^T = \begin{bmatrix} 1 & 1+i & -1 & 1-i \end{bmatrix}.$$

Computing the finite Zak transform,

$$Z_2\mathbf{y} = \begin{bmatrix} 0 & 1 \\ 1 & 1 \end{bmatrix}$$

and

$$Z_2(F(4)\mathbf{y}) = 2\begin{bmatrix} 0 & 1 \\ 1 & i \end{bmatrix}.$$

Theorem 7.4 or its corollary describes the ZS realization of the finite Fourier transform. Up to readdressing, realizing the finite Fourier transform in Zak space requires N multiplications. There is, of course, a cost in inputting and outputting vectors to Zak space.

Corollary 7.2 shows that if the ZS representation of a vector is sparse, then the ZS representation of its finite Fourier transform is sparse. Ignoring the values of the ZS representation but only considering the supports, we can in many applications make use of the relationship between the ZS support of a vector and the ZS support of its finite Fourier transform. Globally these supports are related by the transpose.

Discrete chirps as well as all the signals and sequences constructed in this text have sparse support in Zak space.

7.3 ZS Interchange

Suppose $K = LR$, and $R \ge 1$ is an integer. We will relate Z_L and Z_K by using the decimation-in-time CT FFT matrix factorization [40],

$$F(K) = (F(R) \otimes I_L)\, T_L(K)\, (I_R \otimes F(L))\, P(K, R).$$

This relationship is especially important for dealing with the finite Fourier transform. Theorem 7.4 combined with the result to follow lead to an explicit relationship between $\mathbf{x}$ and $F(N)\mathbf{x}$ on the same Zak space.

Example 7.8 $L = 3$ and $K = 6$. For $\mathbf{x} \in \mathbb{C}^{18}$ write

$$M_3\mathbf{x} = \begin{bmatrix} \mathbf{x}_0 & \mathbf{x}_1 & \mathbf{x}_2 & \mathbf{x}_3 & \mathbf{x}_4 & \mathbf{x}_5 \end{bmatrix},$$

where $\mathbf{x}_k$, $0 \le k < 6$, in $\mathbb{C}^3$ are the stride by 6 decimated components of $\mathbf{x}$. Then

$$P(6,2)M_6\mathbf{x} = \begin{bmatrix} \mathbf{x}_0 & \mathbf{x}_1 & \mathbf{x}_2 \\ \mathbf{x}_3 & \mathbf{x}_4 & \mathbf{x}_5 \end{bmatrix}.$$

Writing

$$Z_3\mathbf{x} = F(3)M_3\mathbf{x} = \begin{bmatrix} X_0 & X_1 & X_2 & X_3 & X_4 & X_5 \end{bmatrix},$$

we have

$$(I_2 \otimes F(3))\, P(6,2)M_6\mathbf{x} = \begin{bmatrix} X_0 & X_1 & X_2 \\ X_3 & X_4 & X_5 \end{bmatrix}.$$

Completing the factorization,

$$Z_6\mathbf{x} = (F(2) \otimes I_3)\, T_3(6) \begin{bmatrix} X_0 & X_1 & X_2 \\ X_3 & X_4 & X_5 \end{bmatrix}.$$

Arguing as in the example we have the following result which relates the $L \times K$ ZS representation of a vector $\mathbf{x} \in \mathbb{C}^N$:

$$Z_L\mathbf{x} = \begin{bmatrix} X_0 & \cdots & X_{K-1} \end{bmatrix}$$

with the $K \times L$ ZS representation of $\mathbf{x}$.

Theorem 7.5

$$Z_K\mathbf{x} = (F(R) \otimes I_L)\, T_L(K) \begin{bmatrix} X_0 & \cdots\cdots & X_{L-1} \\ & \cdot & \\ & \cdot & \\ & \cdot & \\ X_{(R-1)L} & & X_{K-1} \end{bmatrix}.$$

Combining the formulas in Theorems 7.4 and 7.5, we have the following result which relates the $L \times K$ ZS representation of a vector $\mathbf{x} \in \mathbb{C}^N$ with the $L \times K$ ZS representation of $F(N)\mathbf{x}$:

$$Z_L(F(N)\mathbf{x}) = \begin{bmatrix} U_0 & \cdots & U_{K-1} \end{bmatrix}.$$

Corollary 7.4

$$\begin{bmatrix} U_0 & \cdots & U_{L-1} \\ \vdots & & \\ U_{(R-1)L} & \cdots & U_{K-1} \end{bmatrix}$$

$$= KT_L(K)^{-1} \left(F(R)^{-1} \otimes I_L \right) R_K \begin{bmatrix} X_0^T \\ X_1^T D_L(N) \\ \vdots \\ X_{K-1}^T D_L(N)^{K-1} \end{bmatrix}.$$

7.4 Example

$N = LK$ and $\mathbf{y} \in \mathbb{C}^N$. The formulas in Theorem 7.4 and Corollary 7.4 provide ZS realizations of the Fourier transform. If $Z_L\mathbf{y}$ is known, then $Z_K(F(N)\mathbf{y})$ and $Z_L(F(N)\mathbf{y})$ can be computed directly from $Z_L\mathbf{y}$ without reference to $\mathbf{y}$ or $F(N)\mathbf{y}$. The second computation requires $K = LR$, where R is a positive integer.

In this section we derive formulas carrying out these computations when $Z_L\mathbf{y}$ is supported on a single point. These formulas reveal the geometric aspect of the ZS realization of the Fourier transform. In [37, 39] this geometric aspect leads to a decomposition of the indexing set of $L \times K$ Zak space. This decomposition is constructed by the orbits of indexing points determined by powers associated to the ZS realization of $F(N)$. This orbital decomposition leads to a ZS construction of an orthonormal basis diagonalizing $F(N)$ having minimal ZS supports.

Identify $\mathbb{C}^L \times \mathbb{C}^K$ with the space of $L \times K$ complex matrices. Define the mapping $G_K : \mathbb{C}^L \times \mathbb{C}^K \longrightarrow \mathbb{C}^K \times \mathbb{C}^L$ by

$$G_K(Z_L\mathbf{y}) = Z_K(F(N)\mathbf{y}), \quad \mathbf{y} \in \mathbb{C}^N.$$

G_K is a linear isomorphism of $\mathbb{C}^L \times \mathbb{C}^K$ onto $\mathbb{C}^K \times \mathbb{C}^L$ satisfying

$$Z_K^{-1}G_K Z_L = F(N).$$

We call G_K the *realization* of $F(N)$ as a mapping from $\mathbb{C}^L \times \mathbb{C}^K$ onto $\mathbb{C}^K \times \mathbb{C}^L$.

If $L = K$, we have

$$Z_L^{-1}G_L Z_L = F(L^2),$$

and the linear properties of $F(L^2)$ can be identified with the linear properties of G_L. In particular, the eigenvectors of $F(L^2)$ are mapped by Z_L onto the eigenvectors of G_L.

For $0 \le l < L$ and $0 \le k < K$ define $\mathbf{E}(l, k) \in \mathbb{C}^L \times \mathbb{C}^K$ as the vector all of whose components are zero except for the (l, k)-th component whose value is 1. Set $w = e^{2\pi i \frac{1}{N}}$. By Theorem 7.4 we have the following result.

Theorem 7.6 *For $0 \le l < L$ and $0 \le k < K$*

$$G_K(\mathbf{E}(l, k)) = Kw^{lk}\mathbf{E}(l, K - k)^T.$$

T denotes the transpose of $\mathbf{E}(l, K - k)$ viewed as an $L \times K$ matrix.

By Theorem 7.6, G_K maps a vector supported on a single point in $\mathbb{C}^L \times \mathbb{C}^K$ onto a vector supported on a single point in $\mathbb{C}^K \times \mathbb{C}^L$.

Suppose $K = L$. The transformation of supports under G_L are given by the geometric mapping σ of $\mathbb{Z}/L \times \mathbb{Z}/L$ defined by

$$\sigma : (l, k) \longrightarrow (L - k, l), \quad 0 \le l, k < L.$$

Observe $\sigma(0, 0) = (0, 0)$. The *orbit* of a point (l, k) under σ consists of the points

$$(l, k), \quad (L - k, l), \quad (L - l, L - k), \quad (k, L - l),$$

where $\sigma(k, L - l) = (l, k)$. Usually an orbit consists of 4 distinct points, but orbits of size 1 and size 2 can exist.

Example 7.9 $L = K = 3$ and $w = e^{2\pi i \frac{1}{9}}$.

$$G_3(\mathbf{E}(0, 0)) = 3\mathbf{E}(0, 0)$$
$$G_3(\mathbf{E}(0, 1)) = 3\mathbf{E}(2, 0)$$
$$G_3(\mathbf{E}(2, 0)) = 3\mathbf{E}(0, 2)$$
$$G_3(\mathbf{E}(0, 2)) = 3\mathbf{E}(1, 0)$$
$$G_3(\mathbf{E}(1, 0)) = 3\mathbf{E}(0, 1)$$

and

$$G_3(\mathbf{E}(1, 1)) = 3\mathbf{E}(2, 1).$$

We see from the example that $\mathbf{E}(0, 0)$ is an eigenvector with eigenvalue 1 for $\frac{1}{3}G_3$. The space spanned by the orthonormal vectors (disjoint supports)

$$\{\mathbf{E}(0, 1), \ \mathbf{E}(2, 0), \ \mathbf{E}(0, 2), \ \mathbf{E}(1, 0)\}$$

is invariant under $\frac{1}{3}G_3$ and the matrix of the restriction of $\frac{1}{3}G_3$ to this subspace relative to this basis is

$$S_4 = \begin{bmatrix} 0 & 0 & 0 & 1 \\ 1 & 0 & 0 & 0 \\ 0 & 1 & 0 & 0 \\ 0 & 0 & 1 & 0 \end{bmatrix},$$

which can be diagonalized by $F(4)$.

In general we can decompose $\mathbb{Z}/L \times \mathbb{Z}/L$ into disjoint orbits (sizes 1, 2 or 4) under σ. Over each orbit the evaluation vectors, perhaps twiddled with powers of w, are invariant under $\frac{1}{L}G_L$. As supports in an orbit and between orbits are disjoint, we construct an orthonormal basis of $\mathbb{C}^L \times \mathbb{C}^L$ from the twiddled evaluation vectors such that the matrix of $\frac{1}{L}G_L$ with respect to this basis is the matrix direct sum of copies of S_1, S_2 and S_4. This matrix direct sum is diagonalized by a matrix direct sum of copies of $F(1)$, $F(2)$ and $F(4)$, resulting in an orthonormal basis diagonalizing $F(L^2)$. Details can be found in [37, 39].

Define the mapping $G_L : \mathbb{C}^L \times \mathbb{C}^K \longrightarrow \mathbb{C}^L \times \mathbb{C}^K$ by

$$G_L(Z_L \mathbf{y}) = Z_L(F(N)\mathbf{y}), \quad \mathbf{y} \in \mathbb{C}^N.$$

G_L is a linear isomorphism of $\mathbb{C}^L \times \mathbb{C}^K$ such that

$$Z_L^{-1} G_L Z_L = F(N).$$

The linear properties of $F(N)$ can be identified with the linear properties of G_L under Z_L. We call G_L the *realization* of $F(N)$ on $\mathbb{C}^L \times \mathbb{C}^K$. Each factorization of N leads to a new geometric realization of $F(N)$ on a two-dimensional space.

Assume $K = LR$, with R a positive integer. We use Corollary 7.4 to derive a formula for

$$G_L(\mathbf{E}(l,k)), \quad 0 \leq l < L, \, 0 \leq k < K.$$

Combining Theorem 7.6 with Corollary 7.4, where $\mathbf{x} = \mathbf{E}(l,k)$, the right-hand side of the formula in the corollary is

$$KT_L(K)^{-1} \left(F(R)^{-1} \otimes I_L \right) w^{lk} \mathbf{E}(l.K - k)^T.$$

Viewed as a $K \times L$ matrix, $\mathbf{E}(l, K - k)^T$ has all its columns the zero vector except for the l-th column which is $\mathbf{e}^K_{K-k}$. The computation above is complete once we compute

$$KT_L(K)^{-1} \left(F(R)^{-1} \otimes I_L \right) w^{lk} \mathbf{e}^K_{K-k}.$$

Theorem 7.7 *For $0 \leq l < L$ and $0 \leq k < K$*

$$G_L(\mathbf{E}(l,k)) = Lw^{lk} \sum_{t=0}^{R-1} v^{kt} \mathbf{E}(L - s, l + tL),$$

where $s \equiv k \bmod L$ and $v = e^{2\pi i \frac{1}{K}}$.

Proof Write $k = s + rL$, $0 \leq s < L$ and $0 \leq r < R$, and

$$K - k = L - s + (R - r - 1)L.$$

Because

$$\mathbf{e}^K_{K-k} = \mathbf{e}^R_{R-r-1} \otimes \mathbf{e}^L_{L-s},$$

we have

$$\left(F(R)^{-1} \otimes I_L \right) \mathbf{e}^K_{K-k} = \frac{1}{R} \left[u^{(r+1)t} \right]_{0 \leq t < R} \otimes \mathbf{e}^L_{L-s},$$

where $u = e^{2\pi i \frac{1}{R}}$. Completing the computation,

$$KT_L^{-1}(K) \left(F(R)^{-1} \otimes I_L \right) \mathbf{e}^K_{K-k} = L \left[v^{kt} \mathbf{e}^L_{L-s} \right]_{0 \leq t < R},$$

where $v = e^{2\pi i \frac{1}{K}}$ and we have used

$$v^{-(L-s)t} u^{(r+1)t} = v^{kt}.$$

Using the notation of Corollary 7.4,

$$U_{l+tL} = Lw^{lk} v^{kt} \mathbf{e}^L_{L-s}, \quad 0 \leq t < R,$$

and $U_k = \mathbf{0}^L$, otherwise, completing the proof.

Example 7.10 $L = 3$ and $K = 6$.

$$G_3(\mathbf{E}(0,0)) = 3 \begin{bmatrix} 1\,0\,0 & 1\,0\,0 \\ 0\,0\,0 & 0\,0\,0 \\ 0\,0\,0 & 0\,0\,0 \end{bmatrix}$$

$$G_3(\mathbf{E}(0,1)) = 3 \begin{bmatrix} 0\,0\,0 & 0\,0\,0 \\ 0\,0\,0 & 0\,0\,0 \\ 1\,0\,0 & \rho\,0\,0 \end{bmatrix}, \qquad \rho = e^{2\pi i \frac{2}{3}}.$$

View $\mathbb{Z}/L \times \mathbb{Z}/K$ as the indexing set for the coefficients of the vectors in $\mathbb{C}^L \times \mathbb{C}^K$. The support of

$$G_L(\mathbf{E}(l,k)), \qquad s \equiv k \bmod L,$$

consists of the R points in $\mathbb{Z}/L \times \mathbb{Z}/K$

$$(L - s, l), \ (L - s, l + L), \ \ldots, \ (L - s, l + (R - 1)L).$$

We see that for $0 \le l, s < L$, the R vectors

$$G_L(\mathbf{E}(l,s)), \ G_L(\mathbf{E}(l, s + L)), \ \ldots, \ G_L(\mathbf{E}(l, s + (R + 1)L))$$

have the same support. This implies that G_L maps any vector in $\mathbb{C}^L \times \mathbb{C}^K$ supported on the points

$$\{(l, s + tL) : 0 \le t < R\}$$

into a vector in $\mathbb{C}^L \times \mathbb{C}^K$ supported on the points

$$\{(L - s, l + tL) : 0 \le t < R\}.$$

The orbits in this case ($R > 1$) are not orbits of points as in the L^2 case, but orbits of sets of the form described under the geometric map induced by G_L. Details can be found in [37, 39] along with a construction in Zak space of an orthonormal basis diagonalizing $F(N)$, $N = L^2 R$.

8

Zak Space Correlation Formula

The Zak space (ZS) correlation formula expresses the ZS representation of the correlation of two vectors (periodic sequences) in terms of sums of componentwise products of the column vectors of the ZS representations of the two vectors. As we will see in subsequent chapters, this formula is the main tool for studying the correlation properties of discrete chirps and for designing special classes of sequences having good correlation properties. It is a powerful tool for analyzing the correlation properties of vectors whose ZS representations can be factored into products of permutation matrices and diagonal matrices. This includes the discrete chirps, but also an extensively larger set of chirp-like vectors as well.

We begin with the ZS convolution formula and then use the relationship between convolution and correlation to derive the ZS correlation formula.

Throughout this chapter, $N = LK$, where $L > 1$ and $K > 1$ are integers, $Z = Z_L$, $M = M_L$, $S = S_L$, $D = D_L$ and $R = R_L$. $\mathbf{x}$ and $\mathbf{y}$ are vectors in $\mathbb{C}^N$.

The convolution $\mathbf{x} * \mathbf{y}$ can be written as

$$\mathbf{u} = \mathbf{x} * \mathbf{y} = \sum_{n=0}^{N-1} x_n S_N^n \mathbf{y}, \qquad \mathbf{x}, \ \mathbf{y} \in \mathbb{C}^N.$$

Suppose

$$Z\mathbf{x} = \begin{bmatrix} X_0 & \cdots & X_{K-1} \end{bmatrix}, \quad Z\mathbf{y} = \begin{bmatrix} Y_0 & \cdots & Y_{K-1} \end{bmatrix}$$

and

$$Z\mathbf{u} = \begin{bmatrix} U_0 & \cdots & U_{K-1} \end{bmatrix}.$$

As Z is linear,

$$Z\mathbf{u} = \sum_{n=0}^{N-1} x_n Z \left(S_N^n \mathbf{y} \right).$$

Write $n = k + lK$, $0 \le k < K$, $0 \le l < L$. By Theorem 7.1,

$$Z\mathbf{u} = \sum_{k=0}^{K-1} \left(\sum_{l=0}^{L-1} x_{k+lK} D^l \right) Z \left(S_N^k \mathbf{y} \right).$$

M. An et al., *Ideal Sequence Design in Time-Frequency Space*,
DOI 10.1007/978-0-8176-4738-4_8,
© Birkhäuser Boston, a part of Springer Science+Business Media, LLC 2009

As the inner sum is the $L \times L$ diagonal matrix $D\left(X_k\right)$, we have

$$Z\mathbf{u} = \sum_{k=0}^{K-1} X_k Z\left(S_N^k \mathbf{y}\right).$$

The notation $X_k Z\left(S_N^k \mathbf{y}\right)$ means the matrix formed by the componentwise products of the vector X_k with the column vectors of $Z\left(S^k \mathbf{y}\right)$. By Theorem 7.1,

$$X_0 Z\mathbf{y} = \begin{bmatrix} X_0 Y_0 & \cdots & X_0 Y_{K-1} \end{bmatrix},$$

and for $1 \le k < K$,

$$X_k Z\left(S_N^k \mathbf{y}\right) = \begin{bmatrix} D(X_k Y_{K-k}) & \cdots & D(X_k Y_{K-1}) \, X_k Y_0 & \cdots & X_k Y_{K-k+1} \end{bmatrix}.$$

The notation $D(XY)$ means the product of the diagonal matrix $D = D_L$ with the componentwise product vector XY. Summing along the columns, for $0 \le r < K$,

$$U_r = X_0 Y_r + \cdots + X_r Y_0 + D\left(X_{r+1} Y_{K-1} + \cdots + X_{K-1} Y_{r+1}\right),$$

proving the next result.

Theorem 8.1 *For $0 \le r < K$*

$$U_r = \sum_{m=0}^{r} X_m Y_{r-m} + D \sum_{m=r+1}^{K-1} X_m Y_{K-m+r}.$$

The formula in Theorem 8.1 is called the ZS *convolution formula*. It computes the ZS representation of the convolution of two vectors in terms of sums of the componentwise products of the columns of the ZS representation of the two vectors. Except for the matrix multiplication by D, it has the form of convolution on the column vectors of $Z\mathbf{x}$ and $Z\mathbf{y}$. For most of the applications in this text, multiplication by D is a slight problem easily overcome, but its presence, resulting from the noncommuting of $F(N)$ and S_N, is an important consequence of the matrix analog of the uncertainty principle. It will be seen to be a major constraint in analyzing the finite Zak transform of shifted zero-padded vectors, a topic considered in Chapter 14.

Example 8.1 $L = K = 3$.

$$U_0 = X_0 Y_0 + D_3\left(X_1 Y_2 + X_2 Y_1\right),$$
$$U_1 = X_0 Y_1 + X_1 Y_0 + D_3(X_2 Y_2),$$
$$U_2 = X_0 Y_2 + X_1 Y_1 + X_2 Y_0.$$

Example 8.2 $L = K = 3$. Suppose

$$X_0 = E_0, \quad X_1 = E_2, \quad X_2 = E_1$$

and

$$Y_0 = E_2, \quad Y_1 = E_1, \quad Y_2 = E_0,$$

where E_0, E_1 and E_2 form the standard basis in $\mathbb{C}^3$. Then

$$U_0 = D_3 E_1, \quad U_1 = E_2, \quad U_2 = E_0 + E_2,$$

and

$$Z_3 \mathbf{u} = \begin{bmatrix} D_3 E_1 & E_2 & E_0 + E_2 \end{bmatrix}.$$

Setting $w = e^{2\pi i \frac{1}{3}}$, we have

$$M_3 \mathbf{u} = F(3)^{-1} Z_3 \mathbf{u} = \frac{1}{3} \begin{bmatrix} w & 1 & 2 \\ 1 & w & 1+w \\ w^2 & w^2 & 1+w^2 \end{bmatrix}$$

and

$$\mathbf{u}^T = \frac{1}{3} \begin{bmatrix} w & 1 & 2 & 1 & w & 1+w & w^2 & w^2 & 1+w^2 \end{bmatrix}.$$

We see from this example that the ZS convolution formula is easy to apply, when $Z\mathbf{x}$ and $Z\mathbf{y}$ are formed from the vectors in the standard basis.

Set

$$\mathbf{v} = \mathbf{x} \circ \mathbf{y} = \mathbf{x} * R_N \mathbf{y}^*.$$

Because

$$M(R_N \mathbf{y}^*) = \begin{bmatrix} R\mathbf{y}_0^* & S^{-1} R\mathbf{y}_{K-1}^* & \cdots & S^{-1} R\mathbf{y}_1^* \end{bmatrix},$$

and

$$FR = F^* \quad \text{and} \quad FS^{-1}R = D^{-1}F^*,$$

we have

$$Z(R_N \mathbf{y}^*) = FM(R_N \mathbf{y}^*) = \begin{bmatrix} Y_0^* & D^{-1}Y_{K-1}^* & \cdots & D^{-1}Y_1^* \end{bmatrix}.$$

Write the formula in Theorem 8.1 as

$$U_0 = X_0 Y_0 + D \sum_{m=1}^{K-1} X_m Y_{K-m},$$

and for $1 \le r < K$,

$$U_r = \sum_{m=0}^{r-1} X_m Y_{r-m} + X_r Y_0 + D \sum_{m=r+1}^{K-1} X_m Y_{K-m+r}.$$

To compute
$$Z\mathbf{v} = \begin{bmatrix} V_0 \cdots V_{K-1} \end{bmatrix},$$
we replace Y_0 by Y_0^* and Y_m by $D^{-1}Y_{K-m}^*$, $1 \le m < K$, in the formula,
$$V_0 = \sum_{m=0}^{K-1} X_m Y_m^*,$$
and for $1 \le r < K$,
$$V_r = D^{-1} \sum_{m=0}^{r-1} X_m Y_{K+m-k}^* + X_r Y_0^* + \sum_{m=r+1}^{K-1} X_m Y_{m-r}^*,$$
proving the next result, the ZS *correlation formula*.

Theorem 8.2
$$V_0 = \sum_{m=0}^{K-1} X_m Y_m^*$$

and for $1 \le r < K$
$$V_r = D^{-1} \sum_{m=0}^{r-1} X_m Y_{K+m-r}^* + \sum_{m=r}^{K-1} X_m Y_{m-r}^*.$$

Example 8.3 $L = K = 3$.
$$\begin{aligned}
V_0 &= X_0 Y_0^* + X_1 Y_1^* + X_2 Y_2^*, \\
V_1 &= D_3^{-1}(X_0 Y_2^*) + X_1 Y_0^* + X_2 Y_1^*, \\
V_2 &= D_3^{-1}(X_0 Y_1^* + X_1 Y_2^*) + X_2 Y_0^*.
\end{aligned}$$

Example 8.4 $L = K = 3$. Suppose
$$X_0 = E_0, \qquad X_1 = E_2, \qquad X_2 = E_1$$
and
$$Y_0 = E_2, \qquad Y_1 = E_1, \qquad Y_2 = E_0.$$
Then
$$V_0 = \mathbf{0}, \qquad V_1 = D_3^{-1} E_0 + E_2 + E_1 = \mathbf{1}^3, \qquad V_2 = \mathbf{0}$$
and
$$Z\mathbf{v} = \begin{bmatrix} \mathbf{0}\ \mathbf{1}\ \mathbf{0} \end{bmatrix},$$
$$M\mathbf{v} = \begin{bmatrix} \mathbf{0}\ \mathbf{e}_0\ \mathbf{0} \end{bmatrix},$$
implying
$$\mathbf{v}^T = \begin{bmatrix} 0\ 1\ 0\ 0\ 0\ 0\ 0\ 0\ 0 \end{bmatrix}.$$

The same comments made about the ZS convolution formula can be made about the ZS correlation formula.

Theorem 8.2 computes the ZS representation of the correlation, not the correlation directly. It provides an image of the correlation from which the correlation can be computed.

Example 8.5 Setting $T = 50$ and $\gamma = \frac{1}{2}$, $M = 1250$. Set $f = 37.5$. M is divisible by $N = 625 = 25^2$. $u = 2$ and $L = 25$, so $(u, L) = 1$ and $\mathbf{x}_2$ in this case is a unit chirp. Figure 8.1 displays the supports of $Z_{25}\mathbf{x}_2$ and $Z_{25}(\mathbf{x}_2 \circ \mathbf{x}_2)$.

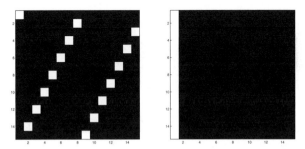

Fig. 8.1. Zak transform supports of a unit chirp and its autocorrelation

Example 8.6 Continuing with Example 8.5, set $f = 33.75$. The resulting sequence $\mathbf{y}$ is not a periodic discrete chirp. Figure 8.2 displays the supports of $Z_{25}\mathbf{y}$ and $Z_{25}(\mathbf{y} \circ \mathbf{y})$ from which we can deduce that $\mathbf{y}$ does not satisfy ideal autocorrelation.

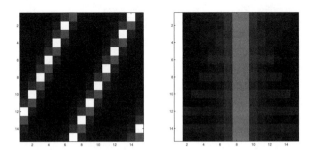

Fig. 8.2. Zak transform supports of a nonperiodic chirp and autocorrelation

Example 8.7 Set $F = F(5)$ and

$$\mathbf{x} = \begin{bmatrix} F_0 \\ F_1 \\ F_2 \\ F_3 \\ F_4 \end{bmatrix}, \quad \mathbf{y} = (S_5 \otimes I_5)\,\mathbf{x} = \begin{bmatrix} F_1 \\ F_2 \\ F_3 \\ F_4 \\ F_0 \end{bmatrix}, \quad \mathbf{z} = (\mu_2 \otimes I_5)\,\mathbf{x} = \begin{bmatrix} F_0 \\ F_2 \\ F_4 \\ F_1 \\ F_3 \end{bmatrix}.$$

Correlations of $\mathbf{x}$, $\mathbf{y}$ and $\mathbf{z}$ were examined in Example 5.8. Figures 8.3 and 8.4 display the supports of ZS representations of the sequences and their cross correlations. Because they satisfy ideal autocorrelation, the Zak transform supports of the autocorrelations are identical. In fact, they are identical to that of $\mathbf{x} \circ \mathbf{y}$.

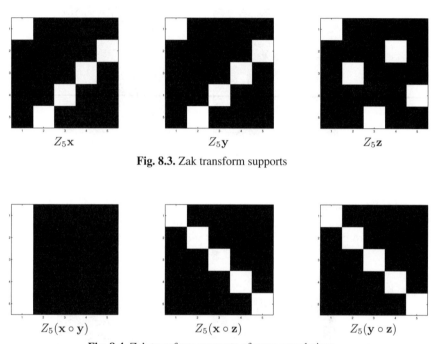

$$Z_5\mathbf{x} \qquad\qquad\qquad Z_5\mathbf{y} \qquad\qquad\qquad Z_5\mathbf{z}$$

Fig. 8.3. Zak transform supports

$$Z_5(\mathbf{x} \circ \mathbf{y}) \qquad\qquad Z_5(\mathbf{x} \circ \mathbf{z}) \qquad\qquad Z_5(\mathbf{y} \circ \mathbf{z})$$

Fig. 8.4. Zak transform supports of cross correlations

Example 8.8 For $L = K = 15$, $\mathbf{x}_2$, $\mathbf{x}_4$ and $\mathbf{x}_7$ with discrete carrier frequencies 0, 0 and $\frac{1}{2}$, respectively, satisfy ideal autocorrelation. $(\mathbf{x}_2, \mathbf{x}_4)$ satisfies ideal correlation while $(\mathbf{x}_2, \mathbf{x}_5)$ does not. The correlations of these sequences were examined in Example 6.2. Figures 8.5 and 8.6 display the supports of ZS representations of the sequences and their cross correlations.

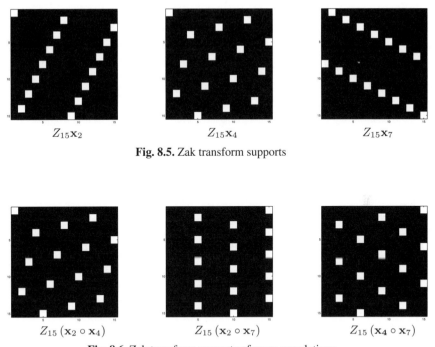

$$Z_{15}\mathbf{x}_2 \qquad Z_{15}\mathbf{x}_4 \qquad Z_{15}\mathbf{x}_7$$

Fig. 8.5. Zak transform supports

$$Z_{15}\left(\mathbf{x}_2 \circ \mathbf{x}_4\right) \qquad Z_{15}\left(\mathbf{x}_2 \circ \mathbf{x}_7\right) \qquad Z_{15}\left(\mathbf{x}_4 \circ \mathbf{x}_7\right)$$

Fig. 8.6. Zak transform supports of cross correlations

9

Zak Space Representation of Chirps

$N = LK = L^2 R$, where $L > 1$ and $R \geq 1$ are integers. $\Lambda = Perm(L)$ and M, Z, S, D, F denote M_L, Z_L, S_L, D_L, $F(L)$.

In this chapter we compute the Zak space (ZS) representation of discrete chirps and study their correlation properties using the ZS correlation formula. The first result shows that the $L \times K$ ZS representation of the unit discrete chirp $\mathbf{x}_u$ in $\mathbb{C}^N$ is a shift of the product of R copies of the unit permutation matrix $E_{-\mu_u}$ with a diagonal matrix. The original use of this result was in analyzing multiple scattering echoes of linear frequency modulated (FM) chirp pulses. The $L \times K$ ZS representation of the critically sampled or subsampled echo is supported on a collection of parallel algebraic lines. The shifts computed from the collection can be identified with radial distances of the targets from the transmitter. For applications, the results of this chapter must be generalized to zero-padded linear FM chirps. This will be done in Chapter 14.

Once the ZS representation of discrete chirps is established, we can apply the ZS correlation formula to study their correlation properties. For simplicity, we restrict this analysis to discrete chirps in $\mathbb{C}^{L^2}$. The discussion reproves the standard correlation results given in Chapter 6, but the method of proof is different. Permutations and permutation matrices play a major role in our approach. The condition

$$(u - v, L) = 1$$

is replaced by a condition on the permutations μ_u and μ_v: the difference

$$\mu_u - \mu_v$$

is a permutation. In the next chapter, we generalize the condition to Λ and show in subsequent chapters that this generalization leads to systematic design procedures for constructing large collections of polyphase sequence sets satisfying pairwise ideal correlation.

We also derive a formula for the ZS representation of the Fourier transform of a unit discrete chirp in the general $N = LK = L^2 R$ case.

Suppose $\mathbf{x}_u$ is a discrete chirp in $\mathbb{C}^N$ having discrete carrier frequency f. Set $v = e^{2\pi i \frac{1}{L}}$. The k-th column of $M\mathbf{x}_u$ is

M. An et al., *Ideal Sequence Design in Time-Frequency Space*,
DOI 10.1007/978-0-8176-4738-4_9,
© Birkhäuser Boston, a part of Springer Science+Business Media, LLC 2009

$$[x_u(k + lK)]_{0 \leq l < L}, \qquad 0 \leq k < K.$$

By Theorem 6.1,

$$x_u(k + lK) = x_u(k)x_u(lK)v^{ukl},$$

where

$$x_u(lK) = e^{\pi i u R l^2} e^{2\pi i f \frac{l}{L}}, \qquad 0 \leq l < L.$$

Using the condition

$$uN + 2f \in 2\mathbb{Z},$$

we show the following result.

Theorem 9.1 *If N is odd and u is even or N is even and uR is even, then*

$$x_u(lK) = v^{fl}, \qquad 0 \leq l < L, \ f \in \mathbb{Z}.$$

If N is odd and u is odd or N is even and uR is odd, then

$$x_u(lK) = v^{f'l}, \qquad 0 \leq l < L, \ f' \in \mathbb{Z},$$

where $f = f' + \frac{L}{2}$.

Proof Suppose N is odd and u is even or N is even and uR is even. Then uR is even and uN is even. uR even implies

$$x_u(lK) = v^{fl}, \qquad 0 \leq l < L.$$

uN even implies $2f \in 2\mathbb{Z}$ and $f \in \mathbb{Z}$, proving the first statement.

Suppose N is odd and u is odd or N is even and uR is odd. Then uR is odd and uN is odd or uN is even. Set $f' = f - \frac{L}{2}$. uR odd implies

$$x_u(lK) = (-1)^l v^{\frac{L}{2}l} v^{f'l} = v^{f'l}, \qquad 0 \leq l < L.$$

uN odd implies L is odd, $2f$ is odd and $f' \in \mathbb{Z}$, completing the proof.

By Theorem 9.1 we can assign to the discrete chirp $\mathbf{x}_u$ a mapping ϕ of $\mathbb{Z}/L$ into itself defined by

$$\phi(l) \equiv -ul - g \mod L,$$

where $g = f$ or $g = f - \frac{L}{2}$ is given by the conditions in Theorem 9.1. ϕ is called the mapping of $\mathbf{x}_u$. ϕ is a permutation, called the *permutation* of $\mathbf{x}_u$, if and only if $\mathbf{x}_u$ is a unit discrete chirp.

Theorem 9.2 *If $\mathbf{x}_u$ is a discrete chirp in $\mathbb{C}^N$, then*

$$Z\mathbf{x}_u = L \left[E_\phi \cdots E_\phi \right] D_K (\mathbf{x}_u),$$

where ϕ is the mapping of $\mathbf{x}_u$, the $L \times L$ matrix E_ϕ is repeated R times and $D_K (\mathbf{x}_u)$ is the $K \times K$ diagonal matrix whose diagonal entries are given by $[x_u(k)]_{0 \leq k < K}$.

Proof The k-th column of $M\mathbf{x}_u$ is

$$x_u(k)D^g D^{uk}\mathbf{1}.$$

Because $FDF^{-1} = S^{-1}$, the k-th column of $Z\mathbf{x}_u$ is

$$Lx_u(k)S^{-g}S^{-uk}\mathbf{e}_0 = Lx_u(k)S^{-g}E_{-\mu_u(k)}, \qquad 0 \le k < K,$$

proving the theorem.

If $\mathbf{x}_u$ is a unit discrete chirp, then ϕ is the permutation of $\mathbf{x}_u$ and E_ϕ is a permutation matrix. In what follows we restrict the discussion to unit discrete chirps. Equally useful results can be derived for arbitrary discrete chirps and more general classes of discrete chirps, but the expression of these results is more complicated.

By Theorem 9.2, if the period of the unit discrete chirp $\mathbf{x}_u$ is $N = L^2$, then the support of its ZS representation is the shifted algebraic line defined by $S^{-g}E_{-\mu_u}$.

Example 9.1 Set $T = 60$ and $\gamma = \frac{1}{2}$ and $f = 600 = 10T$. Figure 9.2 displays the ZS representations of the critically sampled, periodic chirp as surf plots along with its support in the $L \times RL$ image plane. Figure 9.1 displays the ZS representation based on decomposition of $N = 1800$ in a form other than $L \times RL$ for comparison.

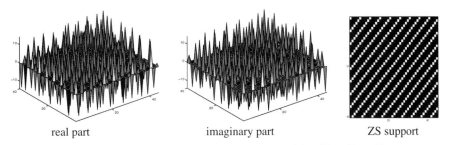

real part   imaginary part   ZS support

Fig. 9.1. Zak transforms of critically sampled chirp, $N = 40 \times 45$

The correlation properties of unit discrete chirps will be investigated using the ZS correlation formula for the case $N = L^2$.

Theorem 9.3 *The unit discrete chirp* $\mathbf{x}_u$ *in* $\mathbb{C}^N$, $N = L^2$, *satisfies ideal autocorrelation.*

Proof The m-th column of $Z\mathbf{x}_u$ is

$$X_m = Lx_u(m)E_{\phi(m)}, \qquad 0 \le m < L,$$

where ϕ is the permutation of $\mathbf{x}_u$. As E_ϕ is a permutation matrix, the componentwise product

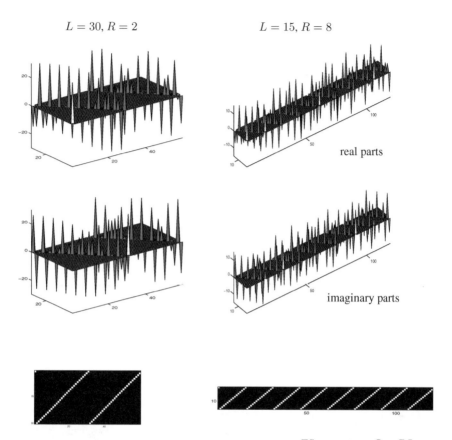

Fig. 9.2. Zak transforms of critically sampled chirp

$$X_m X^*_{m-k}, \qquad 0 \le m < L,$$

is $\mathbf{0}$, unless $k = 0$, in which case

$$X_m X^*_m = L^2 E_{\phi(m)}, \qquad 0 \le m < L.$$

Set $\mathbf{v} = \mathbf{x}_u \circ \mathbf{x}_u$. By the ZS correlation formula the 0-th column of $Z\mathbf{v}$ is

$$L^2 \sum_{m=0}^{L-1} E_{\phi(m)} = L^2 \mathbf{1}$$

and all other columns are the zero vector. Applying F^{-1}, the 0-th column of $M\mathbf{v}$ is

$$L^2 E_0$$

and all other columns are the zero vector, proving the theorem.

Example 9.2 Setting $T = 50$ and $\gamma = \frac{1}{2}$, $M = 1250$. Set $f = 100 = 2T$. γT^2 is divisible by $N = 625 = 25^2$. $u = 2$ and $L = 25$. Because $(u, L) = 1$, x_2 in this case is a unit chirp. Figure 9.3 displays the support of $Z_{25}x_2$. Figure 9.4 displays the autocorrelation of x_2. The imaginary part is zero.

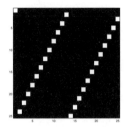

Fig. 9.3. Support of $Z_{25}x_2$

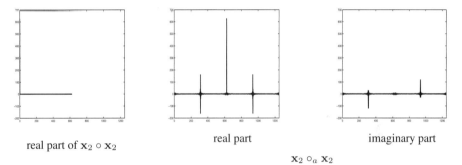

real part of $x_2 \circ x_2$ real part   imaginary part

$x_2 \circ_a x_2$

Fig. 9.4. Autocorrelation of x_2

Suppose x_u and x_v are unit discrete chirps in $\mathbb{C}^N$, $N = L^2$.

$$Zx_u = LE_{\phi_u} D(x_u) \quad \text{and} \quad Zx_v = LE_{\phi_v} D(x_v),$$

where

$$\phi_u(m) = -um - g_u \quad \text{and} \quad \phi_v(m) = -vm - g_v,$$

are the permutations of x_u and x_v. By Theorem 9.3, x_u and x_v satisfy ideal autocorrelation.

Theorem 9.4 *If $u - v \in U_L$, then there exists $\mathbf{c} \in \mathbf{C}_1^L$ and a permutation $\phi_\mathbf{m}$ of $\mathbb{Z}/L$ such that*

$$Z(x_u \circ x_v) = L^2 E_{\phi_u \phi_\mathbf{m}} D(\mathbf{z}),$$

where

$$z_k = c_k x_u(m_k) x_v(m_k - k), \quad 0 \le k < L.$$

Proof Set $\mathbf{w} = \mathbf{x}_u \circ \mathbf{x}_v$. The m-th columns of $Z\mathbf{x}_u$ and $Z\mathbf{x}_v$ are

$$X_m = Lx_u(m)E_{\phi_u(m)}$$

and

$$Y_m = Lx_v(m)E_{\phi_v(m)}.$$

By the ZS correlation formula, the k-th column W_k of $Z\mathbf{w}$ is determined by the componentwise products

$$X_m Y^*_{m-k}, \quad 0 \le m < L, \; m - k \text{ taken modulo } L.$$

For $0 \le k < L$, the componentwise product

$$\left(E_{\phi_u(m)}\right)\left(E_{\phi_v(m-k)}\right)$$

is $\mathbf{0}$, unless m satisfies

$$\phi_u(m) = \phi_v(m-k).$$

We rewrite this condition on m as

$$(v - u)m \equiv vk + g_u - g_v \bmod L.$$

Because $(u - v, L) = 1$, this condition has a unique solution m_k and

$$\phi_u(m_k) = \phi_v(m_k - k).$$

The $m_k, 0 \le k < L$, are distinct mod L and define the permutation $\phi_{\mathbf{m}}$.

It follows that for $0 \le k < L$, the componentwise product $X_m Y^*_{m-k}$ is $\mathbf{0}$, except when $m = m_k$, in which case we have

$$X_{m_k} Y^*_{m_k-k} = L^2 x_u(m_k) x^*_v(m_k - k) E_{\phi_u(m_k)}.$$

By the ZS correlation formula

$$W_k = c_k X_{m_k} Y^*_{m_k-k},$$

where $c_k = w^{\phi_u(m_k)}$ if $m_k < k$ and $c_k = 1$ if $k \le m_k$, completing the proof.

The permutation $\phi_{\mathbf{m}}$ in Theorem 9.4 can be written in terms of the permutations ϕ_u and ϕ_v. From the proof of the theorem,

$$\phi_u \phi_{\mathbf{m}} = \phi_v \left(\phi_{\mathbf{m}} - \phi_0\right),$$

where ϕ_0 is the identity permutation. As a result,

$$\phi_{\mathbf{m}} - \phi_0$$

is a permutation. Because

$$(\phi_v - \phi_u)\phi_{\mathbf{m}} = \phi_v,$$

we have that

$$\phi_0 - \phi_u \phi_v^{-1} = (\phi_v - \phi_u)\,\phi_v^{-1}$$

is a permutation. We can now write

$$\phi_{\mathbf{m}} = \phi_v^{-1}\left(\phi_0 - \phi_u \phi_v^{-1}\right)^{-1}\phi_v\,,$$

proving the following result.

Corollary 9.1 *If $u - v \in U_L$, then*

$$\phi_0 - \phi_u \phi_v^{-1}$$

is a permutation and

$$\phi_{\mathbf{m}} = \phi_v^{-1}\left(\phi_0 - \phi_u \phi_v^{-1}\right)^{-1}\phi_v.$$

The unit discrete chirps $\mathbf{x}_u$ and $\mathbf{x}_v$ satisfy ideal correlation if

$$|w_k| = L, \qquad 0 \le k < N,$$

where $\mathbf{w} = \mathbf{x}_u \circ \mathbf{x}_v$.
By Theorem 9.4,

$$M\left(\mathbf{x}_u \circ \mathbf{x}_v\right) = F^{-1}Z\left(\mathbf{x}_u \circ \mathbf{x}_v\right) = LF_{\phi_u \phi_{\mathbf{m}}}^{*}D(\mathbf{z}).$$

As the components of $\mathbf{z}$ have absolute value 1, we proved the next result.

Theorem 9.5 *If $\mathbf{x}_u$ and $\mathbf{x}_v$ are unit discrete chirps in $\mathbb{C}^N$, $N = L^2$ and $u - v \in U_L$, then $(\mathbf{x}_u, \mathbf{x}_v)$ satisfies ideal correlation.*

Example 9.3 $L = 7$. The unit discrete chirps in $\mathbb{C}^{49}$ are

$$\mathbf{x}_1,\ \mathbf{x}_2,\ \mathbf{x}_3,\ \mathbf{x}_4,\ \mathbf{x}_5,\ \mathbf{x}_6$$

and the pairs of unit discrete chirps in $\mathbb{C}^{49}$ satisfying ideal correlation are the 15 pairs

$$\begin{aligned}
&(\mathbf{x}_1, \mathbf{x}_u), \quad 2 \le u \le 6, \\
&(\mathbf{x}_2, \mathbf{x}_u), \quad 3 \le u \le 6, \\
&(\mathbf{x}_3, \mathbf{x}_u), \quad 4 \le u \le 6, \\
&(\mathbf{x}_4, \mathbf{x}_u), \quad 5 \le u \le 6, \\
&(\mathbf{x}_5, \mathbf{x}_6).
\end{aligned}$$

In general if $L = p$, an odd prime, then there are $p - 1$ unit discrete chirps and $\frac{(p-1)(p-2)}{2}$ ideal correlation discrete chirp pairs in $\mathbb{C}^{p^2}$.

Example 9.4 $L = 9$. The unit discrete chirps in $\mathbb{C}^{81}$ are

$$\mathbf{x}_1,\ \mathbf{x}_2,\ \mathbf{x}_4,\ \mathbf{x}_5,\ \mathbf{x}_7,\ \mathbf{x}_8$$

and the pairs of unit discrete chirps in $\mathbb{C}^{81}$ satisfying ideal correlation are the 8 pairs

$$
\begin{aligned}
(\mathbf{x}_1, \mathbf{x}_u), &\quad u = 2,\ 5,\ 8, \\
(\mathbf{x}_2, \mathbf{x}_u), &\quad u = 4,\ 7, \\
(\mathbf{x}_4, \mathbf{x}_u), &\quad u = 5,\ 8, \\
(\mathbf{x}_7, \mathbf{x}_8). &
\end{aligned}
$$

Example 9.5 Setting $T = 30$, $\gamma_1 = \frac{1}{2}$, $\gamma_2 = \frac{1}{4}$, we have $M_1 = 450$ and $M_2 = 225$. $N = 15^2 = 225$ is a common divisor of M_1 and M_2. Set $\nu_1 = 0$ and $\nu_2 = 1.25$. Then

$$M_1 + 2\nu_1 T = 450 \in 2\mathbb{Z}, \quad \text{and} \quad M_2 + 2\nu_2 T = 300 \in 2\mathbb{Z}.$$

$u = \frac{M_1}{N} = 2$, $v = \frac{M_2}{N} = 1$. As $\mu_1 = \phi_0$, $\mu_2^{-1}\mu_1 = \mu_8$ and

$$\phi_0 - \mu_8 = \begin{pmatrix} 0\ 8\ 1\ 9\ 2\ 10\ 3\ 11\ 4\ 12\ 5\ 13\ 6\ 14\ 7 \end{pmatrix}$$
$$= \mu_8 \in \Lambda.$$

Figures 9.5–9.7 display the correlations of $\mathbf{x}_1$ and $\mathbf{x}_2$. The real parts of $\mathbf{x}_1 \circ \mathbf{x}_1$ and $\mathbf{x}_2 \circ \mathbf{x}_2$ are the delta function at 0 of magnitude $N = 225$, the imaginary parts are zero. Observe that while $\mathbf{x}_1 \circ \mathbf{x}_1 = \mathbf{x}_2 \circ \mathbf{x}_2$, the acyclic autocorrelation can differ significantly. This is an indication of the importance of flexibility in generating a large class of sequence sets with ideal correlation. Figure 9.7 displays the absolute values of $\mathbf{x}_1 \circ_a \mathbf{x}_2$ and $\mathbf{x}_2 \circ_a \mathbf{x}_2$ on the same axis for comparison.

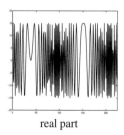
real part

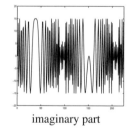
imaginary part

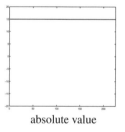
absolute value

Fig. 9.5. Cross correlation, $\mathbf{x}_1 \circ \mathbf{x}_2$

The following theorem uses Theorem 7.4 to derive a formula for the $K \times L$ ZS representation of the unit discrete chirp. We begin with an example.

$\mathbf{x}_1 \circ_a \mathbf{x}_1$

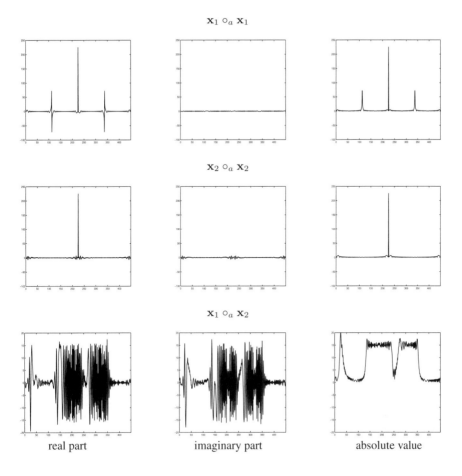

$\mathbf{x}_2 \circ_a \mathbf{x}_2$

$\mathbf{x}_1 \circ_a \mathbf{x}_2$

real part   imaginary part   absolute value

Fig. 9.6. Acyclic correlations

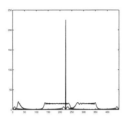

Fig. 9.7. $\mathbf{x}_1 \circ_a \mathbf{x}_2$ and $\mathbf{x}_2 \circ_a \mathbf{x}_2$

Example 9.6 $L = 3$ and $K = 6$. Set

$$D_0(\mathbf{x}) = D\left(\begin{bmatrix} x_0 \\ x_1 \\ x_2 \end{bmatrix}\right), \qquad D_1(\mathbf{x}) = D\left(\begin{bmatrix} x_3 \\ x_4 \\ x_5 \end{bmatrix}\right).$$

Applying Theorem 7.4 to

$$Z_3\mathbf{x}_u = 3\,[E_\phi\ E_\phi]\,D(\mathbf{x}),$$

we have

$$Z_6\mathbf{x}_u = 3\,(F(2) \otimes I_3)\,T_3(6)\begin{bmatrix} E_\phi D_0(\mathbf{x}) \\ E_\phi D_1(\mathbf{x}) \end{bmatrix} = 3\begin{bmatrix} E_\phi D_0(\mathbf{y}) \\ E_\phi D_1(\mathbf{y}) \end{bmatrix},$$

where

$$\begin{bmatrix} D_0(\mathbf{y}) \\ D_1(\mathbf{y}) \end{bmatrix} = (F(2) \otimes I_3)\begin{bmatrix} E_\phi D_0(\mathbf{x}) \\ E_\phi^{-1} D_3(6) E_\phi D_1(\mathbf{x}) \end{bmatrix}.$$

Because E_ϕ is a permutation matrix, $E_\phi^{-1} D_3(6) E_\phi$ is a diagonal matrix,

$$E_\phi^{-1} D_3(6) E_\phi = \begin{bmatrix} w^{\phi(0)} & & \\ & w^{\phi(1)} & \\ & & w^{\phi(2)} \end{bmatrix}, \qquad w = e^{2\pi i \frac{1}{6}},$$

and $D_0(\mathbf{y})$ and $D_1(\mathbf{y})$ are diagonal matrices,

$$D_0(\mathbf{y}) = D\left(\begin{bmatrix} x_0 + w^{\phi(0)} x_3 \\ x_1 + w^{\phi(1)} x_4 \\ x_2 + w^{\phi(2)} x_5 \end{bmatrix}\right), \qquad D_1(\mathbf{y}) = D\left(\begin{bmatrix} x_0 - w^{\phi(0)} x_3 \\ x_1 - w^{\phi(1)} x_4 \\ x_2 - w^{\phi(2)} x_5 \end{bmatrix}\right).$$

For $\mathbf{y} \in \mathbb{C}^N$ define the $L \times L$ diagonal matrix $D_r(\mathbf{y})$ by

$$D_r(\mathbf{y}) = D\left([y_{s+rL}]_{0 \le s < L}\right), \qquad 0 \le r < R.$$

By Theorem 9.2,

$$\begin{aligned} Z_L\mathbf{x}_u &= L\,[E_\phi \quad \cdots \quad E_\phi]\,D(\mathbf{x}) \\ &= L\,[E_\phi D_0(\mathbf{x}) \quad \cdots \quad E_\phi D_{R-1}(\mathbf{x})]. \end{aligned}$$

Theorem 7.4 implies

$$\begin{aligned} Z_K\mathbf{x}_u &= L\,(F(R) \otimes I_L)\,T_L(K)\begin{bmatrix} E_\phi D_0(\mathbf{x}) \\ \vdots \\ E_\phi D_{R-1}(\mathbf{x}) \end{bmatrix} \\ &= L\,(F(R) \otimes I_L)\,T_L(K)\begin{bmatrix} D_0(\mathbf{y})E_\phi \\ \vdots \\ D_{R-1}(\mathbf{y})E_\phi \end{bmatrix}. \end{aligned}$$

Operating by $(F(R) \otimes I_L) T_L(K)$, we have

$$Z_K \mathbf{x}_u = L \begin{bmatrix} D_0(\mathbf{z}) E_\phi \\ \vdots \\ D_{R-1}(\mathbf{z}) E_\phi \end{bmatrix},$$

where

$$\mathbf{z} = (F(R) \otimes I_L) T_L(K) \mathbf{x}',$$

and $\mathbf{x}' = (I_R \otimes E_\phi) \mathbf{x}$. Passing the diagonal matrix factors across E_ϕ, we have the following result.

Theorem 9.6

$$Z_K \mathbf{x}_u = L \left[E_\phi D_r(\mathbf{y}) \right]_{0 \le r < R},$$

where

$$\mathbf{y} = (F(R) \otimes I_L) \left(I_R \otimes E_{\phi^{-1}} \right) T_L(K) \left(I_R \otimes E_\phi \right) \mathbf{x}.$$

As in the example,

$$E_\phi^{-1} D_L^r E_\phi = D \left(\left[w^{r\phi(l)} \right]_{0 \le l < L} \right), \qquad 0 \le r < R,$$

is an $L \times L$ diagonal matrix.

We will now derive formulas for $L \times K$ and $K \times L$ ZS representations of $F(N) \mathbf{x}_u$. A derivation can be given by the formulas in Chapter 6, but we take a ZS approach.

By Theorems 7.4 and 9.2,

$$Z_L (F(N) \mathbf{x}_u) = N R_K \left[x_k E_{\phi(k)}^T D_L(N)^k \right]_{0 \le k < K}$$

$$= N R_K \left[y_k E_{\phi(k)}^T \right]_{0 \le k < K},$$

where

$$y_k = x_k w^{k\phi(k)}, \qquad 0 \le k < K, \;\; w = e^{2\pi i \frac{1}{N}}$$

and k in $\phi(k)$ is taken modulo L. We can write this formula as

$$Z_K (F(N) \mathbf{x}_u) = N R_K \begin{bmatrix} D_0(\mathbf{y}) E_{\phi^{-1}} \\ \vdots \\ D_{R-1}(\mathbf{y}) E_{\phi^{-1}} \end{bmatrix} = N \begin{bmatrix} D_0 (R_K \mathbf{y}) R_L E_{\phi^{-1}} \\ \vdots \\ D_{R-1} (R_K \mathbf{y}) R_L E_{\phi^{-1}} \end{bmatrix}.$$

Passing the diagonal factors across $R_L E_{\phi^{-1}}$, we have the following result.

Theorem 9.7

$$Z_K (F(N) \mathbf{x}_u) = N \begin{bmatrix} R_L E_{\phi^{-1}} D_0(\mathbf{z}) \\ \vdots \\ R_L E_{\phi^{-1}} D_{R-1}(\mathbf{z}) \end{bmatrix},$$

where

$$\mathbf{z} = (E_R \otimes E_\phi R_L) R_K \mathbf{y},$$

and $y_k = x_k w^{k\phi(k)}$, $0 \le k < K$, $w = e^{2\pi i \frac{1}{N}}$.

We want to compute

$$Z_L (F(N)\mathbf{x}_u) = [U_0 \cdots U_{K-1}].$$

By Theorem 7.5 we can compute $Z_L (F(N)\mathbf{x}_u)$ from

$$\begin{bmatrix} U_0 & \cdots & U_{L-1} \\ \vdots & & \\ U_{(R-1)L} & \cdots & U_{K-1} \end{bmatrix} = T_L^{-1}(K) \left(F(R)^{-1} \otimes I_L \right) Z_K (F(N)\mathbf{x}_u).$$

By the proof of the preceding theorem this expression can be written as

$$N \begin{bmatrix} D_0(\mathbf{z}) R_L E_{\phi^{-1}} \\ \vdots \\ D_{R-1}(\mathbf{z}) R_L E_{\phi^{-1}} \end{bmatrix},$$

where

$$\mathbf{z} = T_L^{-1}(K) \left(F(R)^{-1} \otimes I_L \right) R_K \mathbf{y}.$$

Passing the diagonal matrix factors across $R_L E_{\phi^{-1}}$, we have the following result.

Theorem 9.8

$$Z_L (F(N)\mathbf{x}_u) = N \begin{bmatrix} R_L E_{\phi^{-1}} & \cdots & R_L E_{\phi^{-1}} \end{bmatrix} D(\mathbf{w}),$$

where

$$\mathbf{w} = (I_R \otimes E_\phi R_L) T_L^{-1} \left(F(R)^{-1} \otimes I_L \right) R_K \mathbf{y}$$

and

$$y_k = x_k w^{k\phi(k)}, \qquad 0 \le k < K, \ w = e^{2\pi i \frac{1}{N}}.$$

The supports of $Z_L (\mathbf{x}_u)$ and $Z_L (F(N)\mathbf{x}_u)$ are the repeating permutation matrices

$$E_\phi \quad \text{and} \quad R_L E_{\phi^{-1}}$$

R times.

In Chapter 12 we define modulated permutation sequences. Unit discrete chirps and their Fourier transforms are examples by Theorems 9.2 and 9.8. In fact, the arguments used in these theorems prove that the Fourier transform of a modulated permutation sequence is a modulated permutation sequence, up to a scalar multiple.

Example 9.7 $L = 3$ and $R = 2$. $\phi = (0\ 2\ 1)$. The support of $Z_3(\mathbf{x}_u)$ is

$$\begin{bmatrix} 1\,0\,0 & 1\,0\,0 \\ 0\,0\,1 & 0\,0\,1 \\ 0\,1\,0 & 0\,1\,0 \end{bmatrix}$$

and the support of $Z_3(F(18)\mathbf{x}_u)$ is contained in

$$\begin{bmatrix} 1\,0\,0 & 1\,0\,0 \\ 0\,1\,0 & 0\,1\,0 \\ 0\,0\,1 & 0\,0\,1 \end{bmatrix}.$$

Example 9.8 $L = 5$, $R = 2$. $\phi = (0\ 2\ 4\ 1\ 3)$. The support of $Z_5(\mathbf{x}_u)$ is

$$\begin{bmatrix} 1\,0\,0\,0\,0 & 1\,0\,0\,0\,0 \\ 0\,0\,0\,1\,0 & 0\,0\,0\,1\,0 \\ 0\,1\,0\,0\,0 & 0\,1\,0\,0\,0 \\ 0\,0\,0\,0\,1 & 0\,0\,0\,0\,1 \\ 0\,0\,1\,0\,0 & 0\,0\,1\,0\,0 \end{bmatrix}$$

and the support of $Z_5(F(50)\mathbf{x}_u)$ is

$$\begin{bmatrix} 1\,0\,0\,0\,0 & 1\,0\,0\,0\,0 \\ 0\,0\,0\,1\,0 & 0\,0\,0\,1\,0 \\ 0\,1\,0\,0\,0 & 0\,1\,0\,0\,0 \\ 0\,0\,0\,0\,1 & 0\,0\,0\,0\,1 \\ 0\,0\,1\,0\,0 & 0\,0\,1\,0\,0 \end{bmatrix}.$$

10

*-Permutations

$N = L^2$, where $L > 1$ is an integer. $\Lambda = Perm(L)$.

In this chapter we identify a collection of permutations, the *-permutations, and describe some of their properties. We show that *-permutations exist in Λ if and only if L is odd. A key property of a *-permutation is that the fixed point set of every shift of a *-permutation has order one. Future chapters will exploit this property to construct a large number of collections of sequence pairs satisfying ideal correlation.

Consider mappings of $\mathbb{Z}/L$ into itself. Modulus L arithmetic is used for both index set and range set computations.

For $\phi_1, \phi_2 \in \Lambda$, define the mapping $\phi_1 + \phi_2$ of $\mathbb{Z}/L$ into itself by

$$(\phi_1 + \phi_2)(n) = \phi_1(n) + \phi_2(n), \quad n \in \mathbb{Z}/L.$$

$\phi_1 + \phi_2$ is not necessarily a permutation. Denote the identity permutation in Λ by ϕ_0.

A permutation $\phi \in \Lambda$ is called a *-*permutation* if $\phi_0 - \phi$ is a permutation. This definition is motivated by results in [31, 32] which we have repeated in Chapter 6. Because

$$\phi_0 - \mu_u = \mu_{1-u},$$

a unit permutation μ_u is a *-permutation if and only if $1 - u \in U_L$. By Theorem 6.3 a pair of unit discrete chirps $(\mathbf{x}_u, \mathbf{x}_v)$ satisfies ideal correlation if and only if

$$\mu_u^{-1}\mu_v = \mu_{u^{-1}v}$$

is a *-permutation.

The condition

$$1 - u \in U_L$$

is a simple condition for a unit permutation to be a *-permutation. No equivalent condition for an arbitrary permutation exists. To see if an arbitrary permutation ϕ is a *-permutation one has to check that the integers

$$n - \phi(n), \quad n \in \mathbb{Z}/L,$$

are distinct in $\mathbb{Z}/L$.

M. An et al., *Ideal Sequence Design in Time-Frequency Space*,
DOI 10.1007/978-0-8176-4738-4_10,
© Birkhäuser Boston, a part of Springer Science+Business Media, LLC 2009

We begin by showing that ∗-permutations exist if and only if L is odd. The proof is due to Professor Bahman Saffari of the University of Paris and is reproduced here with his permission.

For $\phi \in Perm(L)$ define

$$\Sigma(\phi) = \phi(0) + \phi(1) + \cdots + \phi(L-1).$$

The sum is taken in $\mathbb{Z}/L$. As $\phi(l)$, $0 \leq l < L$, takes on every value in $\mathbb{Z}/L$ exactly once, we have

$$\Sigma(\phi) = \frac{(L-1)L}{2}.$$

In particular, $\Sigma(\phi)$ is independent of ϕ and depends only on L. If L is odd, then $\frac{L-1}{2} \in \mathbb{Z}$ and

$$\Sigma(\phi) \equiv 0 \mod L, \quad L \text{ odd},$$

while if L is even, then $\frac{L}{2} \in \mathbb{Z}$ and

$$\Sigma(\phi) = (L-1)\frac{L}{2} \equiv -\frac{L}{2} \equiv \frac{L}{2}, \mod L, \quad L \text{ even}.$$

Theorem 10.1 *∗-Permutations exist in* $Perm(L)$ *if and only if* L *is odd.*

Proof If L is odd, then the unit permutation μ_2 is a ∗-permutation. Suppose now L is even and $\phi \in Perm(L)$ is a ∗-permutation. By the discussion above, because $\phi_0 - \phi \in Perm(L)$,

$$\Sigma(\phi) = \Sigma(\phi_0 - \phi) \equiv \frac{L}{2} \mod L.$$

However,

$$\Sigma(\phi_0 - \phi) = \Sigma(\phi_0) - \Sigma(\phi) \equiv \frac{L}{2} - \frac{L}{2} \equiv 0 \mod L,$$

a contradiction, completing the proof.

Numerically we have found that the number of ∗-permutations in Λ is significantly larger than the number of unit ∗-permutations in Λ, whenever L is odd and $L > 5$. Moreover, the ratio of the two rapidly increases as L increases and/or as L becomes composite.

Multiplication and addition of unit permutations μ_u, $u \in \mathbb{Z}/L$, are identical to multiplication and addition in $\mathbb{Z}/L$,

$$\mu_u + \mu_v = \mu_{u+v},$$
$$\mu_u \mu_v = \mu_{uv}, \quad u, v \in \mathbb{Z}/L.$$

In particular, the collection of all unit permutations is commutative under multiplication.

General ∗-permutations do not necessarily commute. There is no simple rule for determining if the product of two ∗-permutations is a ∗-permutation. At present computer-aided numerics are necessary, especially in light of the enormous number of ∗-permutations for even relatively small L.

Example 10.1 Tables 10.1–10.3 are examples.

Table 10.1. Unit *-permutations, $L = 7$

	0	1	2	3	4	5	6
μ_2	0	2	4	6	1	3	5
$\phi_0 - \mu_2$	0	6	5	4	3	2	1
μ_3	0	3	6	2	5	1	4
$\phi_0 - \mu_3$	0	5	3	1	6	4	2
μ_4	0	4	1	5	2	6	3
$\phi_0 - \mu_4$	0	4	1	5	2	6	3
μ_5	0	5	3	1	6	4	2
$\phi_0 - \mu_5$	0	3	6	2	5	1	4
μ_6	0	6	5	4	3	2	1
$\phi_0 - \mu_6$	0	2	4	6	1	3	5

In general, if $L = p$, p an odd prime, then the permutations

$$\mu_u, \quad 2 \le u < p,$$

are *-permutations.

Table 10.2. Unit *-permutations, $L = 9$

	0	1	2	3	4	5	6	7	8
μ_2	0	2	4	6	8	1	3	5	7
$\phi_0 - \mu_2$	0	8	7	6	5	4	3	2	1
μ_5	0	5	1	6	2	7	3	8	4
$\phi_0 - \mu_5$	0	5	1	6	2	7	3	8	4
μ_8	0	8	7	6	5	4	3	2	1
$\phi_0 - \mu_8$	0	2	4	6	8	1	3	5	7

Table 10.3. $L = 9$. A permutation which is not a *-permutation

	0	1	2	3	4	5	6	7	8
μ_4	0	4	8	3	7	2	6	1	5
$\phi_0 - \mu_4$	0	6	3	0	6	3	0	6	3

In general if $L = p^2$, p an odd prime, then the unit permutations

$$\mu_u, \quad u \equiv 2, \ldots, p - 1 \bmod p, \quad 0 \le u < p^2 - 1,$$

are *-permutations.

Example 10.2 The following are examples of ∗-permutations that are not unit permutations:

$$L = 7 \quad (0\ 2\ 5\ 1\ 6\ 4\ 3)$$
$$L = 9 \quad (0\ 2\ 1\ 6\ 8\ 7\ 3\ 5\ 4)$$
$$L = 11 \quad (0\ 2\ 4\ 6\ 8\ 10\ 5\ 4\ 3\ 7\ 1)\ .$$

Set Λ^* equal to the collection of all ∗-permutations and Λ_0^* equal to the collection of all unit ∗-permutations.

Example 10.3 $L = 9$.
$$\Lambda_0^*(9) = \{\mu_2,\ \mu_5,\ \mu_8\}\ .$$
$\mu_7 = \mu_2\mu_8$ is not a ∗-permutation and $\mu_5 = \mu_4\mu_8$ is a ∗-permutation.

This example shows that the product of two ∗-permutations is not necessarily a ∗-permutation and that a ∗-permutation can be written as the product of a ∗-permutation and an arbitrary permutation. For $L = p$, p an odd prime, because

$$\Lambda_0^*(p) = \{\mu_u : 1 < u < p\}\ ,$$

the product of two ∗-permutations in $\Lambda_0^*(p)$ is a ∗-permutation if and only if the two are not inverses of each other.

The algebraic structure of ∗-permutations is not completely understood at this time, but there are some general results which play important roles in the construction of sequence sets having good correlation properties.

Suppose $\phi \in \Lambda$. As mappings,

$$(\phi_0 - \phi)\,\phi^{-1} = -\left(\phi_0 - \phi^{-1}\right)\ ,$$

and we see that $\phi_0 - \phi$ is a permutation if and only if $\phi_0 - \phi^{-1}$ is a permutation, proving the following result.

Theorem 10.2 *Suppose $\phi \in \Lambda$. ϕ is a ∗-permutation if and only if ϕ^{-1} is a ∗-permutation.*

Consider the shift permutation $\sigma \in \Lambda$,

$$\sigma = (L - 1\ \ 0\ \ \cdots\ \ L - 2)\ .$$

σ is not a ∗-permutation because

$$\phi_0 - \sigma = (1\ \ 1\ \ \cdots\ \ 1)\ .$$

Example 10.4 ϕ is the ∗-permutation in Example 10.2. $\phi\sigma$ is a ∗-permutation. See Table 10.4.

Table 10.4. Shifting a ∗-permutation, $L = 7$

l	0	1	2	3	4	5	6
ϕ	0	2	5	1	6	4	3
σ	6	0	1	2	3	4	5
$\phi\sigma$	3	0	2	5	1	6	4
$\phi_0 - \phi\sigma$	4	1	0	5	3	6	2

Theorem 10.3 *Suppose $\phi \in \Lambda$. ϕ is a ∗-permutation if and only if $\phi\sigma$ is a ∗-permutation.*

Proof Because

$$\sigma \left(\phi_0 - \phi\sigma \right) (n) = \sigma \left(n - \phi(n-1) \right) = n - 1 - \phi(n-1)$$

and

$$\left(\phi_0 - \phi \right) \sigma(n) = \left(\phi_0 - \phi \right) (n-1) = n - 1 - \phi(n-1),$$

as mappings,

$$\sigma \left(\phi_0 - \phi\sigma \right) = \left(\phi_0 - \phi \right) \sigma.$$

$\phi_0 - \phi\sigma$ is a permutation if and only if $\phi_0 - \phi$ is a permutation, proving the theorem.

By Theorem 10.3, if ϕ is a ∗-permutation, then $\phi\sigma^r, 0 \leq r < L$, is a ∗-permutation. Conversely, if $\phi\sigma^r$ is a ∗-permutation for some $0 \leq r < L$, then ϕ is a ∗-permutation.
Combining Theorems 10.2 and 10.3 with

$$(\sigma\phi)^{-1} = \phi^{-1}\sigma^{-1},$$

we see that ϕ is a ∗-permutation if and only if $\sigma\phi$ is a ∗-permutation. As sets,

$$\sigma^r \Lambda^* = \Lambda^* \sigma^s = \Lambda^*, \quad 0 \leq r, \, s < L.$$

Example 10.5 ϕ is the ∗-permutation in Example 10.2. See Table 10.5.

Table 10.5. $L = 9$. Conjugating a ∗-permutation

l	0	1	2	3	4	5	6	7	8
ϕ	0	2	1	6	8	7	3	5	4
$\phi\mu_4^{-1}$	0	5	7	6	2	4	3	8	1
$\phi_0 - \phi\mu_4^{-1}$	0	5	4	6	2	1	3	8	7
$\mu_4\phi\mu_4^{-1}$	0	2	1	6	8	7	3	5	4
$\phi_0 - \mu_4\phi\mu_4^{-1}$	0	8	1	6	5	7	3	2	4

μ_4 is not a ∗-permutation in $\Lambda(9)$, but $\phi\mu_4^{-1}$ and $\mu_4\phi\mu_4^{-1}$ are ∗-permutations.

Example 10.6 ϕ is the ∗-permutation in Example 10.2.

Table 10.6. Nonclosure of *-permutation.

l	0	1	2	3	4	5	6	7	8
$\mu_5\phi$	0	1	5	3	4	8	6	7	2
$\phi_0 - \mu_4^{-1}\phi$	0	0	6	0	0	6	0	0	6

μ_5 is a *-permutation, but $\mu_5\phi$ is not a *-permutation. See Table 10.6.

Suppose $\phi \in \Lambda$ and μ_u is a unit permutation. As mappings,

$$\phi_0 - \mu_u^{-1}\phi\mu_u = \mu_u^{-1}(\phi_0 - \phi)\mu_u,$$

and $\phi_0 - \mu_u^{-1}\phi\mu_u$ is a permutation if and only if $\phi_0 - \phi$ is a permutation, proving the following result.

Theorem 10.4 *Suppose* $\phi \in \Lambda$ *and* μ_u *is a unit permutation. Then* ϕ *is a *-permutation if and only if* $\mu_u^{-1}\phi\mu_u$ *is a *-permutation.*

Define the *time-reversal permutation* $\rho \in \Lambda$ by

$$\rho(r) = L - r, \qquad 0 \le r < L.$$

$\rho^2 = 1$ and $\rho^{-1} = \rho$. Suppose L is odd. As

$$\phi_0 - \rho = \mu_2,$$

and u_2 is a permutation, ρ is a *-permutation. We will use this result in Chapter 15 to construct real sequences having good correlation properties. For this purpose we need the next result. Suppose $\phi \in \Lambda$ is a *-permutation. As $\rho^2 = 1$,

$$\phi_0 - \rho^{-1}\phi\rho = (\rho - \rho^{-1}\phi)\rho$$

and

$$\rho - \rho^{-1}\phi = -(\phi_0 - \phi),$$

then

$$\rho^{-1}\phi\rho = \rho\phi\rho$$

is a *-permutation. We summarize these results in the following theorem.

Theorem 10.5 *The time-reversal permutation* ρ *is a *-permutation.* $\rho^2 = \rho$ *and for any *-permutation* ϕ, $\rho\phi\rho$ *is a *-permutation.*

The following tensor product construction generates *-permutations for composite size L.

Set $\Lambda_1 = Perm(L_1)$, $\Lambda_2 = Perm(L_2)$ and $\Lambda = Perm(L_1L_2)$. The tensor product of two permutation matrices is a permutation matrix.

$$E_\gamma = E_\phi \otimes E_\delta, \qquad \phi \in \Lambda_1, \ \delta \in \Lambda_2, \ \text{and} \ \gamma \in \Lambda,$$

where γ is defined by

$$\gamma(s + rL_2) = \phi(r)L_2 + \delta(s), \qquad 0 \le r < L_1, \ 0 \le s < L_2.$$

We call γ the *tensor product* of ϕ and δ and write $\gamma = \phi \otimes \delta$.

Theorem 10.6 *If $\phi \in \Lambda_1^*$ and $\delta \in \Lambda_2^*$, then $\gamma = \phi \otimes \delta \in \Lambda^*$.*

Proof   We have to show that the set of integers modulo $L_1 L_2$

$$\{s + rL_2 - \gamma(s + rL_2), \quad 0 \leq r < L_1,\ 0 \leq s < L_2\}$$

has order $L_1 L_2$. Suppose

$$s + rL_2 - \gamma(s + rL_2) \equiv s' + r'L_2 - \gamma(s' + r'L_2) \mod L_1 L_2,$$

where $0 \leq s,\ s' < L_2$ and $0 \leq r,\ r' < L_1$. Then

$$s - \delta(s) \equiv s' - \delta(s') \mod L_2.$$

Because δ is a $*$-permutation in Λ_2, $s = s'$. Then

$$r - \phi(r) \equiv r' - \phi(r') \mod L_1.$$

Because ϕ is a $*$-permutation in Λ_1, $r = r'$, completing the proof.

Example 10.7 $L_1 = L_2 = 3$ and $\phi = \delta = \mu_2 \in \Lambda_1$, where

$$\mu_2(n) \equiv 2n \mod 3, \quad 0 \leq n < 3.$$

$$\gamma = \phi \otimes \delta = \begin{pmatrix} 0\ 2\ 1\ 6\ 8\ 7\ 3\ 5\ 4 \end{pmatrix}$$

is a $*$-permutation, but not a unit $*$-permutation.

In general, the tensor product of two unit $*$-permutations is a $*$-permutation, but not a unit permutation.

10.1 Fixed Points

In Chapter 11 we assign to each permutation $\gamma \in \Lambda$ a sequence e_γ in $\mathbb{C}^N$, called the permutation sequence defined by γ. As we will see, the correlation properties of permutation sequences are closely related to the fixed points in $\mathbb{Z}/L$ of shifts of γ.

Suppose $\gamma \in \Lambda$. For $0 \leq k < L$ define the subset $\Delta_k(\gamma)$ of $\mathbb{Z}/L$ by

$$\Delta_k(\gamma) = \{0 \leq m < L : m = \gamma(m - k)\},$$

where $m - k$ is taken modulo L. $\Delta_k(\gamma)$ is the fixed point set of the permutation $\gamma \sigma^k$.

If γ is a $*$-permutation, then by Theorem 10.2, $\phi_0 - \gamma^{-1}$ is a permutation. For each $0 \leq k < L$, there exists a unique $0 \leq m < L$ such that

$$\left(\phi_0 - \gamma^{-1} \right) m = k.$$

m is the unique element in $\Delta_k(\gamma)$ and $|\Delta_k(\gamma)| = 1$.

Conversely, if for $0 \leq k < L$, $\Delta_k(\gamma)$ has order 1, then $\phi_0 - \gamma^{-1}$ is a permutation, proving the next result.

Theorem 10.7 *Suppose $\gamma \in \Lambda$. γ is a *-permutation if and only if*

$$|\Delta_k(\gamma)| = 1, \qquad 0 \le k < L.$$

*-Permutations play a role similar to that played by shift sequences of interleaved sequences in sequence design over finite fields [18, 20, 21]. It should be observed that the space of sequence design in finite fields is the space determined by M while the space of sequence design in this work is Zak space.

Suppose γ is a *-permutation. Because γ^{-1} is a *-permutation,

$$\phi_{\mathbf{m}} = \left(\phi_0 - \gamma^{-1}\right)^{-1}$$

is a permutation, where $\phi_m = (m_0 \ \ m_1 \ \ \cdots \ \ m_{L-1})$. Multiplying on the left by $\phi_0 - \gamma^{-1}$ we have

$$\left(\phi_0 - \gamma^{-1}\right) \phi_{\mathbf{m}} = \phi_{\mathbf{m}} - \gamma^{-1} \phi_{\mathbf{m}} = \phi_0$$

and

$$\phi_{\mathbf{m}} - \phi_0 = \gamma^{-1} \phi_{\mathbf{m}}.$$

In particular, $\phi_{\mathbf{m}}$ is a *-permutation. Evaluating each side at $k, 0 \le k < L$, we have

$$m_k - k = \gamma^{-1}(m_k)$$

and

$$m_k = \gamma(m_k - k), \qquad 0 \le k < L,$$

proving the following result.

Theorem 10.8 *If γ is a *-permutation, then*

$$\phi_{\mathbf{m}} = \left(\phi_0 - \gamma^{-1}\right)^{-1}$$

*is a *-permutation and*

$$\Delta_k(\gamma) = \{m_k\}, \qquad 0 \le k < L.$$

Set $\Delta_k = \Delta_k(\gamma)$ in the following examples.

Example 10.8 $L = 9$ and $\gamma = \mu_2$.

$$\Delta_k = \{2k\}, \qquad 0 \le k < 9,$$

where $2k$ is taken modulo 9.

Example 10.9 $L = 9$ and $\gamma = \mu_5$.

$$\Delta_0 = \{0\} \text{ and } \Delta_k = \{9 - k\}, \quad 1 \leq k < 9.$$

Example 10.10 $L = 9$ and $\gamma = \mu_4$.

$$\Delta_0 = \{0, 3, 6\}, \quad \Delta_3 = \{1, 4, 7\}, \quad \Delta_6 = \{2, 5, 8\}.$$

The remaining Δ_k are the empty set.

Example 10.11 $L = 9$ and

$$\gamma = \begin{pmatrix} 0 & 2 & 8 & 5 & 7 & 3 & 1 & 6 & 4 \end{pmatrix}$$

$$\Delta_0 = \{0\},$$

$$\Delta_1 = \{2\}, \quad \Delta_2 = \{5\}, \quad \Delta_3 = \{7\}, \quad \Delta_4 = \{1\},$$
$$\Delta_5 = \{4\}, \quad \Delta_6 = \{8\}, \quad \Delta_7 = \{3\}, \quad \Delta_8 = \{6\},$$

10.2 Permutation Matrices

The orders of the fixed point sets

$$\Delta_k(\gamma), \quad 0 \leq k < L,$$

will be related to the traces of the matrices

$$E_\gamma S^k, \quad 0 \leq k < L.$$

Set

$$E_\gamma = \begin{bmatrix} E_{\gamma(0)} \cdots E_{\gamma(L-1)} \end{bmatrix}.$$

The trace of E_γ is the sum of the diagonal entries of E_γ which is equal to the number of times

$$\gamma(m) = m, \quad 0 \leq m < L.$$

Example 10.12 $L = 5$ and $\gamma = \mu_2$.

$$E_\gamma = \begin{bmatrix} 1 & 0 & 0 & 0 & 0 \\ 0 & 0 & 0 & 1 & 0 \\ 0 & 1 & 0 & 0 & 0 \\ 0 & 0 & 0 & 0 & 1 \\ 0 & 0 & 1 & 0 & 0 \end{bmatrix}$$

and

$$Tr E_\gamma = \Delta_0(\gamma) = 1.$$

Because the permutation $\gamma\sigma^k$, $0 \le k < L$, determines the permutation matrix

$$E_\gamma S_L^{-k},$$

the trace of $E_\gamma S_L^{-k}$ is equal to the number of times

$$\gamma\sigma^k(m) = \gamma(m - k) = m, \quad 0 \le m < L,$$

which is the order of $\Delta_k(\gamma)$. In general we have

$$\mathrm{Tr}\left(E_\gamma S_L^{-k}\right) = |\Delta_k(\gamma)|, \quad 0 \le k < L.$$

Theorem 10.9 *For $\gamma \in \Lambda$,*

$$L = \sum_{k=0}^{L-1} |\Delta_k(\gamma)|.$$

Proof By linearity of Tr

$$\sum_{k=0}^{L-1} \mathrm{Tr}\left(E_\gamma S_L^{-k}\right) = \mathrm{Tr}\left(E_\gamma \sum_{k=0}^{L-1} S_L^{-k}\right) = \mathrm{Tr}\left(E_\gamma I(L, L)\right),$$

where $I(L, L)$ is the $L \times L$ matrix of all ones. The theorem follows from

$$\mathrm{Tr}\left(E_\gamma I(L, L)\right) = \mathrm{Tr}(I(L, L)) = L.$$

Corollary 10.1 *For $\gamma \in \Lambda$,*

$$|\Delta_k(\gamma)| > 0, \quad 0 \le k < L,$$

if and only if

$$|\Delta_k(\gamma)| = 1, \quad 0 \le k < L.$$

11

Permutation Sequences

$N = L^2$, where $L > 1$ is an integer. Unless otherwise specified, matrices and construction are taken relative to L.

In this chapter we begin sequence design directly in $L \times L$ Zak space. The resulting sequences will have period $N = L^2$. The general case $N = LK = L^2R$ will be considered in Chapter 12. By Theorem 9.2, a unit discrete chirp of period $N = L^2$ has Zak space (ZS) representation

$$LE_\phi D(\mathbf{x}), \quad \phi \in \Lambda, \ \mathbf{x} \in \mathbb{C}_1^L.$$

One of the key features of sequence design in Zak space is that we can decouple the permutation matrix factor from the diagonal matrix factor in the design strategy. In this chapter we study the role of the permutation matrix factor in sequence design. This study is based on the fact that the componentwise product of any two distinct columns of a permutation matrix is the zero vector. Placed inside the ZS correlation formula this fact leads to sparse and well-behaved ZS representations of autocorrelation and cross correlation. For cross correlation this structure of ZS representations is closely linked to the fixed point sets corresponding to shifts of permutations.

For $\phi \in \Lambda$, define $\mathbf{e}_\phi \in \mathbb{C}^N$ by

$$Z\mathbf{e}_\phi = E_\phi.$$

$\mathbf{e}_\phi$ is called a *permutation sequence*. Set $\mathbf{v} = \mathbf{e}_\phi \circ \mathbf{e}_\phi$. By the ZS correlation formula,

$$V_0 = \sum_{m=0}^{L-1} E_{\phi(m)} E_{\phi(m)} = \sum_{m=0}^{L-1} E_{\phi(m)} = \mathbf{1},$$

and

$$V_k = D^{-1} \sum_{m=0}^{k-1} E_{\phi(m)} E_{\phi(m-k)} + \sum_{m=k}^{L-1} E_{\phi(m)} E_{\phi(m-k)}, \quad 1 \leq k < L,$$

where $m - k$ is taken modulo L. Because

M. An et al., *Ideal Sequence Design in Time-Frequency Space*,
DOI 10.1007/978-0-8176-4738-4_11,
© Birkhäuser Boston, a part of Springer Science+Business Media, LLC 2009

$$E_{\phi(m)} E_{\phi(m-k)} = 0, \qquad 1 \le k < L,$$

we have $V_k = 0, 1 \le k < L$, and

$$Z\mathbf{v} = \begin{bmatrix} 1 & 0 & \cdots & 0 \end{bmatrix}.$$

Then

$$M\mathbf{v} = F^{-1} Z\mathbf{v} = \begin{bmatrix} E_0 & 0 & \cdots & 0 \end{bmatrix},$$

where F is the $L \times L$ Fourier transform matrix and $\mathbf{v} = \mathbf{e}_0^N$, proving the following result [12, 23].

Theorem 11.1 *If $\phi \in \Lambda$, then $\mathbf{e}_\phi$ satisfies the ideal autocorrelation property.*

Example 11.1 For $L = 15$, Figures 11.1 and 11.2 display $\mathbf{e}_{\mu_2}$ and its acyclic autocorrelation.

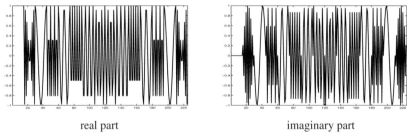

real part   imaginary part

Fig. 11.1. Permutation sequence

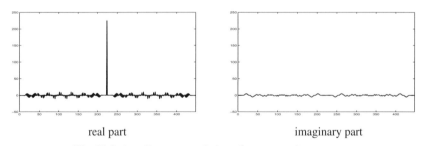

real part   imaginary part

Fig. 11.2. Acyclic autocorrelation of a permutation sequence

Example 11.2 For $L = 3$ and $\phi = (0 \ 1 \ 2)$,

$$Z\mathbf{e}_\phi = I_3, \qquad M\mathbf{e}_\phi = \frac{1}{3} F^*(3)$$

$$\mathbf{e}_\phi^T = \frac{1}{3} \begin{bmatrix} 1 & 1 & 1 & 1 & v^2 & v & 1 & v & v^2 \end{bmatrix}, \qquad v = e^{2\pi i \frac{1}{3}}.$$

Example 11.3 For $L = 3$ and $\phi = (1 \; 2 \; 0)$,

$$Z\mathbf{e}_\phi = \begin{bmatrix} E_1 & E_2 & E_0 \end{bmatrix}, \qquad M\mathbf{e}_\phi = \frac{1}{3} \begin{bmatrix} F_1^* & F_2^* & F_0^* \end{bmatrix}$$

and

$$\mathbf{e}_\phi^T = \frac{1}{3} \begin{bmatrix} 1 & 1 & 1 & v^2 & v & 1 & v & v^2 & 1 \end{bmatrix}, \qquad v = e^{2\pi i \frac{1}{3}}.$$

11.1 Computation in Zak Space

One of the motivations for using ZS representation is that processing can take place directly in Zak space. ZS processing is especially advantageous if some of the sequences in the processing have sparse and well-structured ZS representations.

In this section we compute correlation, convolution and multiplication in Zak space where one of the sequences in the computation is a permutation sequence. The method easily extends to the modulated permutation sequences defined in Chapter 12, including discrete chirps.

Consider $\phi \in \Lambda$ and $\mathbf{y} \in \mathbb{C}^N$. Write

$$Z\mathbf{y} = \begin{bmatrix} Y_0 & \cdots & Y_{L-1} \end{bmatrix}$$

and

$$Y_s = [Y_s(r)]_{0 \leq r < L}, \qquad 0 \leq s < L.$$

We begin by computing the ZS representation of the cross correlation $\mathbf{v} = \mathbf{e}_\phi \circ \mathbf{y}$. As we will see, this involves permuting the components of the column vectors of $Z\mathbf{y}$.

The componentwise product

$$E_{\phi(r)} Y_s = Y_s(\phi(r)) E_{\phi(r)}$$

has 0 in all components except for the $\phi(r)$-th component. The $\phi(r)$-th component is equal to the $\phi(r)$-th component of Y_s. Placing this result in the ZS correlation formula we have

$$V_0 = \sum_{r=0}^{L-1} Y_r^*(\phi(r)) E_{\phi(r)},$$

and for $1 \leq s < L$

$$V_s = \sum_{r=0}^{L-1} Y_{r-s}^*(\phi(r)) v^{-\phi_s(r)}, \qquad v = e^{2\pi i \frac{1}{L}},$$

where $r - s$ is taken modulo L and

$$\phi_s(r) = \begin{cases} \phi(r), & 0 \leq r < s, \\ 0, & s \leq r < L. \end{cases}$$

Equating components,

$$V_0(\phi(r)) = Y_r^*(\phi(r)), \quad 0 \leq r < L,$$

and for $1 \leq s < L$

$$V_s(\phi(r)) = Y_{r-s}^*(\phi(r))v^{-\phi_s(r)}, \quad 0 \leq r < L.$$

As $E_\phi^{-1}\mathbf{x} = [x(\phi(r))]_{0 \leq r < L}$, we have

$$E_\phi^{-1}V_0 = [Y_r^*(\phi(r))]_{0 \leq r < L},$$

and for $1 \leq s < L$

$$E_\phi^{-1}V_s = \left[Y_{r-s}^*(\phi(r))v^{-\phi_s(r)}\right]_{0 \leq r < L},$$

proving the following result.

Theorem 11.2

$$E_\phi^{-1}Z\left(\mathbf{e}_\phi \circ \mathbf{y}\right) = \left[Y_{r-s}^*(\phi(r))v^{-\phi_s(r)}\right]_{0 \leq r, \, s < L}, \quad v = e^{2\pi i \frac{1}{L}},$$

where

$$\phi_s(r) = \begin{cases} \phi(r), & 0 \leq r < s, \\ 0, & s \leq r < L. \end{cases}$$

Example 11.4 $L = 3$. $\phi = (0 \ 2 \ 1)$ and $\mathbf{y} \in \mathbb{C}^9$.

$$E_\phi^{-1}Z_3\left(\mathbf{e}_\phi \circ \mathbf{y}\right) = \begin{bmatrix} y_0^*(0) & y_2^*(0) & y_1^*(0) \\ y_1^*(2) & y_0^*(2) & y_2^*(2)v^{-2} \\ y_2^*(1) & y_1^*(1) & y_0^*(1) \end{bmatrix}.$$

Example 11.5 $L = 5$. $\phi = (0 \ 2 \ 4 \ 1 \ 3)$ and $\mathbf{y} \in \mathbb{C}^{25}$.

$$E_\phi^{-1}Z_5\left(\mathbf{e}_\phi \circ \mathbf{y}\right) = \begin{bmatrix} y_0^*(0) & y_4^*(0) & y_3^*(0) & y_2^*(0) & y_1^*(0) \\ y_1^*(2) & y_0^*(2) & y_4^*(2)v^{-2} & y_3^*(2)v^{-2} & y_2^*(2)v^{-2} \\ y_2^*(4) & y_1^*(4) & y_0^*(4) & y_4^*(4)v^{-4} & y_3^*(4)v^{-4} \\ y_3^*(1) & y_2^*(1) & y_1^*(1) & y_0^*(1) & y_4^*(1)v^{-1} \\ y_4^*(3) & y_3^*(3) & y_2^*(3) & y_1^*(3) & y_0^*(3) \end{bmatrix}.$$

By Theorem 11.2 the ZS representation of the cross correlation $\mathbf{v} = \mathbf{e}_\phi \circ \mathbf{y}$ is formed by permuting the coefficients of $Z\mathbf{y}$ followed by $\frac{L^2-L}{2}$ multiplications of the permuted coefficients by powers of v determined by $\phi_s(r)$.

Direct and inverse finite Zak transforms are required to transfer $\mathbf{y}$ to Zak space and to transfer $Z(\mathbf{e}_\phi \circ \mathbf{y})$ back to sequence space at a cost of $2L^2 \log L$ complex add-multiply using the Cooley-Tukey fast Fourier transform algorithm. The total cost of computing $\mathbf{e}_\phi \circ \mathbf{y}$ is $2L^2 \log L$ additions and $2L^2 \log L + \frac{L^2-L}{2}$ multiplications.

ZS computation of cross correlation provides an *image* of the cross correlation that can be directly analyzed. Suppose, for example,

$$\mathbf{y} = a\mathbf{e}_\phi + bS_N\mathbf{e}_\phi, \qquad a,\ b \in \mathbb{C}.$$

Then

$$E_\phi Z\mathbf{y} = aE_\phi$$

and

$$Z\left(S_N\mathbf{e}_\phi\right)^* Z\mathbf{y} = bE_\phi S^{-1},$$

from which we can compute a and b.

The convolution $\mathbf{w} = \mathbf{e}_\phi * \mathbf{y}$ can be handled in much the same way by the ZS convolution formula. For $0 \leq s < L$

$$W_s = \sum_{r=0}^{s} Y_{s-r}(\phi(r))E_{\phi(r)} + \sum_{r=s+1}^{L-1} Y_{s-r}(\phi(r))v^{\phi(r)}E_{\phi(r)},$$

where $r - s$ is taken modulo L. The components of W_s are then

$$W_s(\phi(r)) = \begin{cases} Y_{s-r}(\phi(r)), & 0 \leq r \leq s, \\ Y_{s-r}(\phi(r))v^{\phi(r)}, & s < r < L, \end{cases}$$

proving the next result.

Theorem 11.3

$$E_\phi^{-1} Z\left(\mathbf{e}_\phi * \mathbf{y}\right) = \left[Y_{s-r}(\phi(r))v^{\phi_s(r)}\right]_{0 \leq r,\ s < L}, \qquad v = e^{2\pi i \frac{1}{L}},$$

where

$$\phi_s(r) = \begin{cases} 0, & 0 \leq r \leq s, \\ \phi(r), & s < r < L. \end{cases}$$

Example 11.6 $L = 3$. $\phi = (0\ 2\ 1)$ and $\mathbf{y} \in \mathbb{C}^9$.

$$E_\phi^{-1} Z_3\left(\mathbf{e}_\phi * \mathbf{y}\right) = \begin{bmatrix} Y_0(0) & Y_1(0) & Y_2(0) \\ Y_2(2) & Y_0(2) & Y_1(2) \\ Y_1(1) & Y_2(1) & Y_0(1) \end{bmatrix}.$$

Example 11.7 $L = 5$. $\phi = (0 \; 2 \; 4 \; 1 \; 3)$ and $\mathbf{y} \in \mathbb{C}^5$.

$$E_\phi^{-1} Z_5 \left(\mathbf{e}_\phi * \mathbf{y} \right) = \begin{bmatrix} Y_0(0) & Y_1(0) & Y_2(0) & Y_3(0) & Y_4(0) \\ Y_4(2)v^2 & Y_0(2) & Y_1(2) & Y_2(2) & Y_3(2) \\ Y_3(4)v^4 & Y_4(4)v^4 & Y_0(4) & Y_1(4) & Y_2(4) \\ Y_2(1)v & Y_3(1)v & Y_4(1)v & Y_0(1) & Y_1(1) \\ Y_1(3)v^3 & Y_2(3)v^3 & Y_3(3)v^3 & Y_4(3)v^3 & Y_0(3) \end{bmatrix}.$$

Consider multiplication by a permutation sequence, $\mathbf{u} = \mathbf{e}_\phi \mathbf{y}$, $\mathbf{y} \in \mathbb{C}^N$, $\phi \in \Lambda$. Then

$$u(n) = e_\phi(n)y(n), \qquad 0 \le n < N.$$

As

$$M\mathbf{e}_\phi = \frac{1}{L} \left[F_{\phi(0)}^* \quad \cdots \quad F_{\phi(L-1)}^* \right],$$

we have

$$M\mathbf{u} = \frac{1}{L} \left[D^{-\phi(0)}\mathbf{y}_0 \quad \cdots \quad D^{-\phi(L-1)}\mathbf{y}_{L-1} \right],$$

where $\mathbf{y}_s$, $0 \le s < L$, are the decimated components of $\mathbf{y}$. Then

$$Z\mathbf{u} = \frac{1}{L} \left[S^{\phi(0)}Y_0 \quad \cdots \quad S^{\phi(L-1)}Y_{L-1} \right],$$

proving the next result.

Theorem 11.4

$$Z \left(\mathbf{e}_\phi \mathbf{y} \right) = \frac{1}{L} \left[S^{\phi(0)}Y_0 \quad \cdots \quad S^{\phi(L-1)}Y_{L-1} \right].$$

Example 11.8 $L = 3$. $\phi = (0 \; 2 \; 1)$.

$$Z \left(\mathbf{e}_\phi \mathbf{y} \right) = \frac{1}{3} \left[Y_0 \quad S^2 Y_1 \quad S Y_2 \right].$$

The ZS representation of the finite Fourier transform of a permutation sequence $\mathbf{e}_\phi$, $\phi \in \Lambda$, in $\mathbb{C}^N$ will be computed using Theorem 7.4. Because

$$Z\mathbf{e}_\phi = E_\phi,$$

we have

$$Z \left(F(N)\mathbf{e}_\phi \right) = R \begin{bmatrix} E_{\phi(0)}^T \\ E_{\phi(0)}^T D_L(N) \\ \vdots \\ E_{\phi(0)}^T D_L(N)^{L-1} \end{bmatrix}.$$

Using

$$E_{\phi(l)}^T D_L(N)^l = w^{l\phi(l)} E_{\phi(l)}^T, \qquad w = e^{2\pi i \frac{1}{N}},$$

we can write

$$Z\left(F(N)\mathbf{e}_\phi\right) = RD\left(\left[w^{l\phi(l)}\right]_{0 \le l < L}\right) E_{\phi^{-1}}$$

$$= RE_{\phi^{-1}}E_\phi D\left(\left[w^{l\phi(l)}\right]_{0 \le l < L}\right) E_{\phi^{-1}}$$

$$= RE_{\phi^{-1}}D\left(\left[w^{l\phi^{-1}(l)}\right]_{0 \le l < L}\right),$$

proving the following result.

Theorem 11.5 *If $\mathbf{e}_\phi$, $\phi \in \Lambda$, is a permutation sequence in $\mathbb{C}^N$, then*

$$Z\left(F(N)\mathbf{e}_\phi\right) = E_{\rho\phi^{-1}}D(\mathbf{y}),$$

where $\rho \in \Lambda$ is defined by $\rho = (0 \quad L-1 \quad \cdots \quad 1)$ and $\mathbf{y} \in \mathbb{C}^L$ is defined by

$$y_l = w^{l\phi^{-1}(l)}, \qquad 0 \le l < L, \quad w = e^{2\pi i \frac{1}{N}}.$$

In particular, the finite Fourier transform of a permutation sequence in $\mathbb{C}^N$, $N = L^2$, is a modulated permutation sequence in $\mathbb{C}^N$.

11.2 Ideal Correlation of Permutation Sequences

In this section we study the correlation properties of permutation sequence pairs in $\mathbb{C}^N$

$$(\mathbf{e}_\phi, \mathbf{e}_\delta), \qquad \phi, \delta \in \Lambda.$$

$*$-Permutations play a major role in this study. We show that each $*$-permutation determines a large collection of permutation sequence pairs satisfying ideal correlation. In Chapter 13 we use this result to design sequence sets satisfying pairwise ideal correlation.

Suppose $\phi, \delta \in \Lambda$ and $\gamma = \phi^{-1}\delta$ and set $\mathbf{w} = \mathbf{e}_\phi \circ \mathbf{e}_\delta$. The main result of this section is that γ is a $*$-permutation if and only if $(\mathbf{e}_\phi, \mathbf{e}_\delta)$ satisfies ideal correlation. We first describe the ZS representation of $\mathbf{w}$. γ is not necessarily a $*$-permutation. Set

$$\Delta_k^0(\gamma) = \{m \in \Delta_k(\gamma) : 0 \le m < k\}$$

and

$$\Delta_k^1(\gamma) = \{m \in \Delta_k(\gamma) : k \le m < L\}.$$

These sets are introduced to match the ZS correlation formula. In particular, $\Delta_k(\gamma)$ is the disjoint union of $\Delta_k^0(\gamma)$ and $\Delta_k^1(\gamma)$.

Theorem 11.6

$$W_0 = \sum_{m \in \Delta_0(\gamma)} E_{\phi(m)}$$

and for $1 \leq k < L$

$$W_k = D^{-1} \sum_{m \in \Delta_k^0(\gamma)} E_{\phi(m)} + \sum_{m \in \Delta_k^1(\gamma)} E_{\phi(m)}.$$

Proof   By the ZS correlation formula,

$$W_0 = \sum_{m=0}^{L-1} E_{\phi(m)} E_{\delta(m)}$$

and for $1 \leq k < L$

$$W_k = D^{-1} \sum_{m=0}^{k-1} E_{\phi(m)} E_{\delta(m-k)} + \sum_{m=k}^{L-1} E_{\phi(m)} E_{\delta(m-k)}.$$

Because for $0 \leq k < L$

$$E_{\phi(m)} E_{\delta(m-k)} = \begin{cases} E_{\phi(m)}, & \text{whenever } m \in \Delta_k(\gamma), \\ \mathbf{0}, & \text{otherwise,} \end{cases} \qquad 0 \leq m < L,$$

we have

$$W_0 = \sum_{m \in \Delta_0(\gamma)} E_{\phi(m)}$$

and for $1 \leq k < L$

$$W_k = D^{-1} \sum_{m \in \Delta_k(\gamma)} E_{\phi(m)} + \sum_{m \in \Delta_k^1(\gamma)} E_{\phi(m)},$$

completing the proof.

Suppose γ is a $*$-permutation and

$$\phi_{\mathbf{m}} = \left(\phi_0 - \gamma^{-1} \right)^{-1}.$$

Theorem 10.8 implies

$$\Delta_k(\gamma) = \{m_k\}, \qquad 0 \leq k < L.$$

By Theorem 11.6

$$W_0 = E_{\phi(m_0)}$$

and for $1 \leq k < L$

$$W_k = c_k E_{\phi(m_k)},$$

where

$$c_k = \begin{cases} v^{-\phi(m_k)}, & 0 \le m_k < k, \\ 1, & k \le m_k < L, \end{cases} \quad v = e^{2\pi i \frac{1}{L}},$$

proving the following result.

Theorem 11.7 *If γ is a $*$-permutation, then*

$$Z\left(\mathbf{e}_\phi \circ \mathbf{e}_\delta\right) = E_{\phi\phi_\mathbf{m}} D(\mathbf{c}),$$

where $\phi_\mathbf{m} = \left(\phi_0 - \gamma^{-1}\right)^{-1}$ and

$$c_k = \begin{cases} v^{-\phi(m_k)}, & 0 \le m_k < k, \\ 1, & k \le m_k < L, \end{cases} \quad v = e^{2\pi i \frac{1}{L}}.$$

Up to modulation by $D(\mathbf{c})$, the ZS representation of the cross correlation of two permutation sequences has the same form as the ZS representations of the permutation sequences themselves. In fact, if γ is a $*$-permutation, then we can write the formula in Theorem 11.7 as

$$Z\left(\mathbf{e}_\phi \circ \mathbf{e}_\delta\right) = Z\left(\mathbf{e}_{\phi\phi_\mathbf{m}}\right) D(\mathbf{c}).$$

Example 11.9 Table 11.1 lists the values of c for $L = 5$. $\phi = \gamma_2$, $\delta = \gamma_3$, $\gamma = \gamma_4$ *and*

$$\phi_\mathbf{m} = \gamma_3 \text{ and } \phi\phi_\mathbf{m} = \gamma_4.$$

Table 11.1. Table for $\mathbf{c}$

k	0	1	2	3	4
m_k	0	3	1	4	2
c_k	1	1	v^{-2}	1	v^{-4}

Then

$$Z\left(\mathbf{e}_\phi \circ \mathbf{e}_\delta\right) = E_{\gamma_4} D(\mathbf{c})$$

and

$$\left(\mathbf{e}_\phi \circ \mathbf{e}_\delta\right)^T =$$

$$\begin{bmatrix} 1 & 1 & v^3 & 1 & 1 & 1 & v & 1 & v^3 & 1 & 1 & v^2 & v^2 & v & v^3 & 1 & v^3 & v^4 & v^4 & v^2 & 1 & v^4 & v & v^2 & v \end{bmatrix}.$$

If γ is a $*$-permutation, then by Theorem 11.7

$$M\mathbf{w} = \frac{1}{L}F^*_{\phi\phi_m}D(\mathbf{c}),$$

where F is the $L \times L$ Fourier transform matrix. This implies

$$|w_n| = \frac{1}{L}, \qquad 0 \le n < N,$$

and $(\mathbf{e}_\phi, \mathbf{e}_\delta)$ satisfies ideal correlation.

Conversely, if $(\mathbf{e}_\phi, \mathbf{e}_\delta)$ satisfies ideal correlation, then every component of $\mathbf{w}$ is nonzero and no column of $Z\mathbf{w}$ is $\mathbf{0}$. By Theorem 11.7 and Corollary 10.1,

$$|\Delta_k(\gamma)| = 1, \qquad 0 \le k < L.$$

Theorem 10.7 implies γ is a $*$-permutation, proving the following theorem.

Theorem 11.8 γ *is a $*$-permutation if and only if* $(\mathbf{e}_\phi, \mathbf{e}_\delta)$ *satisfies ideal correlation.*

From the proof of Theorem 11.8 we have the following corollary.

Corollary 11.1 $(\mathbf{e}_\phi, \mathbf{e}_\delta)$ *satisfies ideal correlation if and only if* $\mathbf{e}_\phi \circ \mathbf{e}_\delta$ *has no zero components.*

By Theorem 11.8, each $\gamma \in \Lambda^*$ determines a collection of permutation sequence pairs

$$\{(\mathbf{e}_\phi, \mathbf{e}_{\phi\gamma}) : \phi \in \Lambda\}$$

satisfying ideal correlation. We call $\gamma \in \Lambda^*$ the *root* of the collection.

Every permutation sequence pair satisfying ideal correlation is contained in at least one of these collections. The collection of permutation sequence pairs having root γ has order

$$\frac{L!}{2} \text{ if } \gamma^2 = \phi_0$$

and

$$L! \text{ if } \gamma^2 \ne \phi_0.$$

Moreover, γ and γ^{-1} determine the same collection of permutation sequence pairs.

Example 11.10 For $L = 31$, set $\gamma = \mu_2$. For any $\phi \in \Lambda$, the pair $(\mathbf{e}_\phi, \mathbf{e}_{\phi\gamma})$ satisfies ideal correlation. While the correlation properties are independent of ϕ, the shapes of the sequences and acyclic correlation properties can vary. Figures 11.3–11.5 display the sequence pairs and their acyclic correlations for some ϕ. For comparison, acyclic autocorrelations and cross correlations are plotted on the same axis.

$\mathbf{e}_{\phi_0}$

$\mathbf{e}_{\phi_0\mu_2}$

Acyclic correlations

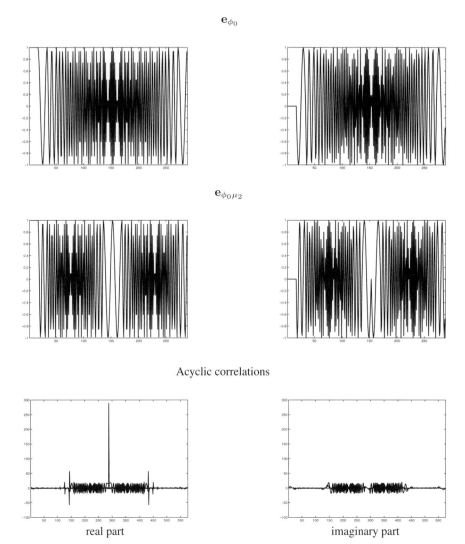

real part   imaginary part

Fig. 11.3. $\phi = \phi_0$

$$\mathbf{e}_{\phi}$$

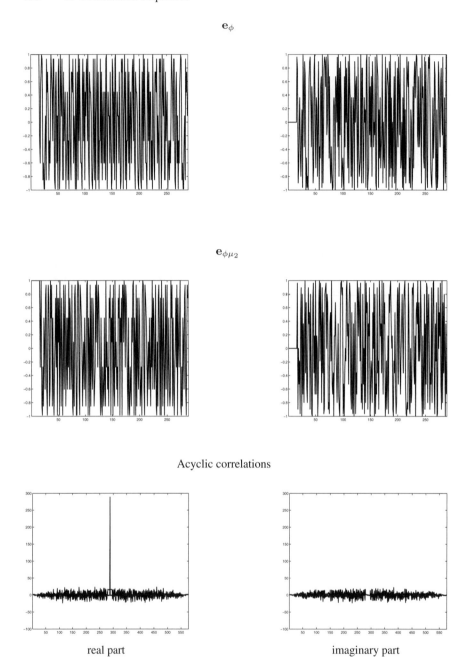

$$\mathbf{e}_{\phi\mu_2}$$

Acyclic correlations

real part   imaginary part

Fig. 11.4. $\phi = (12 \ \ 11 \ \ 3 \ \ 2 \ \ 0 \ \ 16 \ \ 7 \ \ 6 \ \ 8 \ \ 9 \ \ 5 \ \ 13 \ \ 10 \ \ 1 \ \ 14 \ \ 15 \ \ 4)$

$$\mathbf{e}_\phi$$

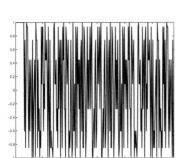 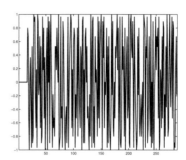

$$\mathbf{e}_{\phi\mu_2}$$

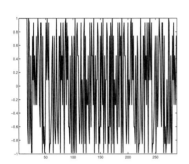 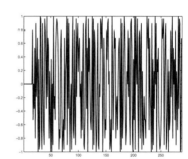

Acyclic correlations

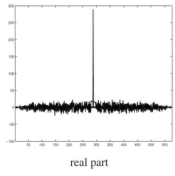 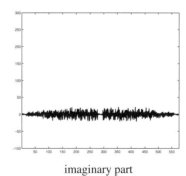

real part   imaginary part

Fig. 11.5. $\phi = (11\ \ 15\ \ 7\ \ 4\ \ 0\ \ 16\ \ 1\ \ 5\ \ 10\ \ 3\ \ 6\ \ 13\ \ 14\ \ 12\ \ 8\ \ 2\ \ 9)$

11.3 Relaxing the *-Permutation Condition

Ideal correlation is a direct consequence of the *-permutation property. By relaxing this property we give up ideal correlation, but increase the number of available permutation pairs. The goal is to relax the ∗-permutation property in such a way that the loss of ideal correlation is controlled as to both magnitude and position. L in this section is not necessarily odd.

Suppose $\gamma \in \Lambda$. For $0 \le k < L$,

$$|\Delta_k(\gamma)| = \left|\left\{0 \le n < L : (\phi_0 - \gamma)^{-1} n = k\right\}\right|.$$

γ is a ∗-permutation if and only if for $0 \le k < L$

$$|\Delta_k(\gamma)| = 1.$$

We now consider the correlation consequences of having the order of some $\Delta_k(\gamma)$ greater than 1. By Theorem 10.9,

$$\sum_{k=0}^{L-1} |\Delta_k(\gamma)| = L,$$

so if some of the sets $\Delta_k(\gamma)$ have order greater than one, then others must be empty.

Suppose ϕ and δ are in Λ and $\gamma = \phi^{-1}\delta$. Set $v = e^{2\pi i \frac{1}{L}}$,

$$r_k = |\Delta_k(\gamma)|, \qquad 0 \le k < L,$$

and

$$r = \max\{r_k : 0 \le k < L\}.$$

By Theorem 10.8, the decimated components of $\mathbf{y} = \mathbf{e}_\phi \circ \mathbf{e}_\delta$ are given by

$$\mathbf{y}_0 = \frac{1}{L} \sum_{m \in \Delta_0(\gamma)} F^*_{\phi(m)}$$

and for $1 \le k < L$

$$\mathbf{y}_k = \frac{1}{L} \sum_{m \in \Delta_k^0(\gamma)} v^{-\phi(m)} F^*_{\phi(m)} + \frac{1}{L} \sum_{m \in \Delta_k^1(\gamma)} F^*_{\phi(m)},$$

where F_l is the l-th column vector of F. Write

$$\mathbf{y}_k = [y_k(l)]_{0 \le l < L}.$$

Theorem 11.9 *For $0 \le l < L$*

$$|y_k(l)| \le \frac{r_k}{L}, \qquad 0 \le l < L,$$

and

$$|y(n)| \le \frac{r}{L}, \qquad 0 \le n < L^2.$$

By Theorem 11.9, if $r > 1$, then $(\mathbf{e}_\phi, \mathbf{e}_\delta)$ does not satisfy ideal correlation, but the coefficients of the correlation $\mathbf{e}_\phi \circ \mathbf{e}_\delta$ are bounded by $\frac{r}{L}$.

In the following examples,

$$\mathbf{y} = \mathbf{e}_{\phi_0} \circ \mathbf{e}_\gamma.$$

γ is not a *-permutation. Note how the values of the components of $\mathbf{y}$ and their positions vary. Also, we can take L to be even since we do not require a *-permutation.

Example 11.11 $L = 9$, $F = F(9) = [F_0 \quad \cdots \quad F_8]$ and $w = e^{2\pi i \frac{1}{9}}$. See Table 11.2.

Table 11.2.

k	0	1	2	3	4	5	6	7	8
γ^{-1}	0	7	5	3	1	8	6	4	2
$|\Delta_k(\gamma)|$	3	0	0	3	0	0	3	0	0

$$\mathbf{y}_0 = \frac{1}{9}\left(F_0^{'*} + F_3^{'*} + F_6^{'*}\right),$$

$$\mathbf{y}_3 = \frac{1}{9}\left(w^{-1}F_1^* + F_4^* + F_7^*\right),$$

and

$$\mathbf{y}_6 = \frac{1}{9}\left(w^{-2}F_2^* + w^{-5}F_5^* + F_8^*\right).$$

The remaining decimated components of $\mathbf{y}$ vanish.

Example 11.12 $L = 9$. See Table 11.3.

Table 11.3.

k	0	1	2	3	4	5	6	7	8
γ^{-1}	3	4	5	6	7	8	0	1	2
$|\Delta_k(\gamma)|$	0	0	0	0	0	0	9	0	0

Because $\gamma^{-1} = \sigma^6$, with σ the shift permutation, the only nonzero decimated component is $\mathbf{y}_6$.

Example 11.13 $L = 9$. See Table 11.4.

Table 11.4.

k	0	1	2	3	4	5	6	7	8
γ^{-1}	3	4	8	5	6	7	0	1	2
$|\Delta_k(\gamma)|$	0	0	0	1	0	0	5	3	0

$$\mathbf{y}_3 = \frac{1}{9} w^{-2} F_2^*,$$

$$\mathbf{y}_6 = \frac{1}{9} \left(F_0^* + w^{-1} F_1^* + F_6^* + F_7^* + F_8^* \right),$$

and

$$\mathbf{y}_7 = \frac{1}{9} \left(w^{-3} F_3^* + w^{-4} F_4^* + w^{-5} F_5^* \right).$$

The remaining decimated components of **y** vanish.

Example 11.14 $L = 8$. See Table 11.5.

Table 11.5.

k	0	1	2	3	4	5	6	7
γ^{-1}	0	3	6	1	4	7	2	5
$\lvert \Delta_k(\gamma) \rvert$	2	0	2	0	2	0	2	0

Example 11.15 $L = 10$. See Table 11.6.

Table 11.6.

k	0	1	2	3	4	5	6	7	8	9
γ^{-1}	5	2	4	6	8	0	7	9	1	3
$\lvert \Delta_k(\gamma) \rvert$	0	0	0	0	0	2	2	2	2	2

Examples 11.14 and 11.15 can be generalized to the case $L = 2M$, where M is an odd integer. As in the examples, $r = 2$ and

$$\lvert y(n) \rvert \le \frac{2}{L}, \quad 0 \le n < L^2.$$

Example 11.16 For $L = 42$, define γ^{-1} by

$$\gamma^{-1}(0) = 21, \quad \gamma^{-1}(k) = 2k, \quad 1 \le k \le 20, \quad \gamma^{-1}(k + 21) = \gamma^{-1}(k) + 21,$$

$$k \le 0 \le 20.$$

$$\lvert \Delta_k(\gamma) \rvert = \begin{cases} 0, & 0 \le k < 21 \\ 2, & 21 \le k < 42. \end{cases}$$

Figure 11.6 displays the correlations of $\mathbf{e}_{\phi_0}$ and $\mathbf{e}_\gamma$. For comparison, the autocorrelations and the cross correlation are plotted on the same axis.

$$\mathbf{e}_{\phi_0} \circ \mathbf{e}_{\gamma}$$

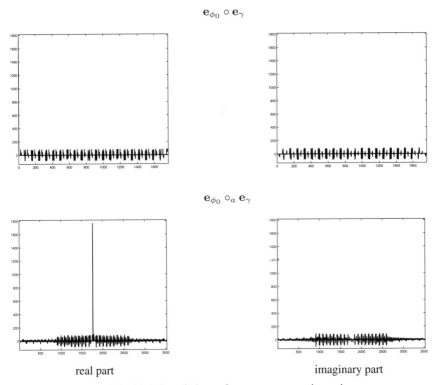

$$\mathbf{e}_{\phi_0} \circ_a \mathbf{e}_{\gamma}$$

real part   imaginary part

Fig. 11.6. Correlations of a non-∗-permutation pair

12

Modulation

$N = L^2 R$, where $L > 1$, $R \geq 1$ are integers. $\Lambda = \Lambda(L)$. $M = M_L$, $Z = Z_L$ and $F = F(L)$.

The Zak space (ZS) representation of a unit discrete chirp in $\mathbb{C}^N$, up to a scalar multiple by L, is a shift of the product of R copies of a permutation matrix with a diagonal matrix having diagonal entries with absolute value 1. ZS methods permit decoupling of the permutations and diagonal matrices. In preceding chapters we used these methods to study and generalize the permutation matrix part. We now bring back the diagonal matrix and study modulation of permutation sequences. Modulation always refers to diagonal matrix multiplication in Zak space.

We first extend the framework developed for permutation sequences in $\mathbb{C}^N$, $N = L^2$, to modulated permutation sequences in $\mathbb{C}^N$. The main result shows that modulation does not affect correlation properties.

Then, beginning with Section 12.2 we define permutation sequences in $\mathbb{C}^N$, $N = LK = L^2 R$, $R > 1$. Permutation sequences in $\mathbb{C}^N$ do not satisfy ideal auto-correlation. Conditions on modulation are established for ideal autocorrelation and pairwise ideal cross correlation.

12.1 Modulation $N = L^2$

Throughout this section $N = L^2$, $\phi, \delta \in \Lambda$ and $\gamma = \phi^{-1}\delta$.

For $\mathbf{x} \in \mathbb{C}_1^L$, define $\mathbf{x}_\phi$ in $\mathbb{C}^N$ by

$$Z\mathbf{x}_\phi = E_\phi D(\mathbf{x}).$$

We call $\mathbf{x}_\phi$ a *modulated* permutation sequence in $\mathbb{C}^N$ and say that $\mathbf{x}$ *modulates* $\mathbf{e}_\phi$ [31, 36]. As $\mathbf{x} \in \mathbb{C}_1^L$ and

$$M\mathbf{x}_\phi = \frac{1}{L} F_\phi^* D(\mathbf{x}),$$

we have $|\mathbf{x}_\phi(n)| = \frac{1}{L}$ and $||\mathbf{x}_\phi|| = 1$.

Theorem 12.1 $\mathbf{x}_\phi$ *satisfies ideal autocorrelation.*

M. An et al., *Ideal Sequence Design in Time-Frequency Space*,
DOI 10.1007/978-0-8176-4738-4_12,
© Birkhäuser Boston, a part of Springer Science+Business Media, LLC 2009

Proof Set $\mathbf{v} = \mathbf{x}_\phi \circ \mathbf{x}_\phi$. By the ZS correlation formula,

$$V_0 = \sum_{m=0}^{L-1} E_{\phi(m)} = \mathbf{1}, \qquad \mathbf{1} = \mathbf{1}^L,$$

and for $1 \le k < L$

$$V_k = D^{-1} \sum_{m=0}^{k-1} x_m x_{m-k}^* E_{\phi(m)} E_{\phi(m-k)} + \sum_{m=k}^{L-1} x_m x_{m-k}^* E_{\phi(m)} E_{\phi(m-k)},$$

where $m - k$ is taken modulo L. Because for $1 \le k < L$,

$$E_{\phi(m)} E_{\phi(m-k)} = \mathbf{0}, \qquad 0 \le m < L,$$

we have

$$V_k = \mathbf{0}, \qquad 1 \le k < L,$$

$$Z\mathbf{v} = [\mathbf{1}\ \mathbf{0}\ \cdots\ \mathbf{0}], \quad \text{and} \quad M\mathbf{v} = [E_0\ \mathbf{0}\ \cdots\ \mathbf{0}],$$

completing the proof.

By Theorem 12.1, modulation does not change the ideal autocorrelation of the permutation sequence $\mathbf{e}_\phi$ in $\mathbb{C}^N$.

The next theorem describes the ZS representation of the cross correlation of a pair of modulated sequences $\mathbf{x}_\phi$ and $\mathbf{y}_\delta$ in $\mathbb{C}^N$. Set $\mathbf{w} = \mathbf{x}_\phi \circ \mathbf{y}_\delta$.

Theorem 12.2

$$W_0 = \sum_{m \in \Delta_0(\gamma)} x_m y_m^* E_{\phi(m)}$$

and for $1 \le k < L$

$$W_k = D^{-1} \sum_{m \in \Delta_k^0(\gamma)} x_m y_{m-k}^* E_{\phi(m)} + \sum_{m \in \Delta_k^1(\gamma)} x_m y_{m-k}^* E_{\phi(m)},$$

where $m - k$ *is taken modulo L.*

Proof By the ZS correlation formula

$$W_0 = \sum_{m=0}^{L-1} x_m y_m^* E_{\phi(m)} E_{\delta(m)}.$$

Because $E_{\phi(m)} E_{\delta(m)} = \mathbf{0}$, unless $m \in \Delta_0(\gamma)$ and $E_{\phi(m)} E_{\gamma(m)} = E_{\phi(m)}$, whenever $m \in \Delta_0(\gamma)$, we have

$$W_0 = \sum_{m \in \Delta_0(\delta)} x_m y_m^* E_{\phi(m)}.$$

For $1 \le k < L$

$$W_k = D^{-1} \sum_{m=0}^{k-1} x_m y_{m-k}^* E_{\phi(m)} E_{\delta(m-k)} + \sum_{m=k}^{L-1} x_m y_{m-k}^* E_{\phi(m)} E_{\delta(m-k)}.$$

Because $E_{\phi(m)} E_{\delta(m-k)} = 0$, unless $m \in \Delta_k(\gamma)$ and $E_{\phi(m)} E_{\delta(m-k)} = E_{\phi(m)}$, whenever $m \in \Delta_k(\gamma)$, we have

$$W_k = D^{-1} \sum_{m \in \Delta_k^0(\gamma)} x_m y_{m-k}^* E_{\phi(m)} + \sum_{m \in \Delta_k^1(\gamma)} x_m y_{m-k}^* E_{\phi(m)},$$

completing the proof of the theorem.

Suppose $\gamma = \phi^{-1}\delta$ is a $*$-permutation and $\phi_{\mathbf{m}} = (\phi_0 - \gamma^{-1})^{-1}$. Then

$$\Delta_k(\gamma) = \{m_k\}, \quad 0 \le k < L.$$

By Theorem 12.2,

$$W_0 = x_{m_0} y_{m_0}^* E_{\phi(m_0)},$$

and for $1 \le k < L$,

$$W_k = c_k x_{m_k} y_{m_k - k}^* E_{\phi(m_k)},$$

where

$$c_k = \begin{cases} v^{-\phi(m_k)}, & m_k \in \Delta_k^0 \\ 1, & \text{otherwise} \end{cases}.$$

This discussion proves the next result.

Theorem 12.3 *If γ is a $*$-permutation and $\phi_{\mathbf{m}} = (\phi_0 - \gamma^{-1})^{-1}$, then*

$$Z(\mathbf{x}_\phi \circ \mathbf{y}_\delta) = E_{\phi\phi_{\mathbf{m}}} D(\mathbf{z}),$$

where $z_k = c_k x_{m_k} y_{m_k - k}^*$, $0 \le k < L$.

By Theorem 12.3 if γ is a $*$-permutation, then

$$M\mathbf{w} = \frac{1}{L} F_{\phi\phi_{\mathbf{m}}}^* D(\mathbf{z}).$$

Because $\mathbf{z} \in \mathbb{C}_1^L$, we have

$$|w_n| = \frac{1}{L}, \quad 0 \le n < N,$$

proving the next result, in one direction.

Theorem 12.4 *γ is a $*$-permutation if and only if $(\mathbf{x}_\phi, \mathbf{y}_\delta)$ satisfies ideal correlation.*

Proof Suppose $(\mathbf{x}_\phi, \mathbf{y}_\delta)$ satisfies ideal correlation. Then

$$|w_n| = \frac{1}{L}, \quad 0 \le n < N.$$

In particular, no component of $\mathbf{w}$ vanishes, implying no column vector of $Z\mathbf{w}$ is $\mathbf{0}$. By Theorem 12.2,

$$\Delta_k(\delta) > 0, \quad 0 \le k < L,$$

and by Corollary 10.1 and Theorem 12.4 γ is a $*$-permutation, completing the proof of the theorem.

From the proof of Theorem 12.4, γ is a $*$-permutation if and only if no component of $\mathbf{x}_\phi \circ \mathbf{y}_\delta$ vanishes.

As with ideal autocorrelation, modulation does not change the correlation properties of a pair of permutation sequences in $\mathbb{C}^N$.

Example 12.1 Figure 12.1 plots the modulating vector $\mathbf{x}$ in

$$Z_{25}\mathbf{x}_2 = \mu_{24} D(\mathbf{x})$$

for the unit discrete chirp.

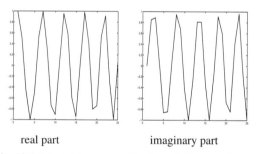

real part   imaginary part

Fig. 12.1. Modulating vector for the unit discrete chirp $\mathbf{x}_2$

Modulation does not affect the correlation. However, it can significantly affect the shape of the sequence and the acyclic correlation. Figure 12.2 plots the modified modulating vector $\mathbf{y}$. It is obtained by interpolating the first 5 values of $\mathbf{x}$. Acyclic autocorrelation of the resulting signal is compared with the acyclic autocorrelation of $\mathbf{x}_2$. Note that the sidelobe of the acyclic autocorrelation of the modified sequence is considerably less.

Modified modulating vector **y**

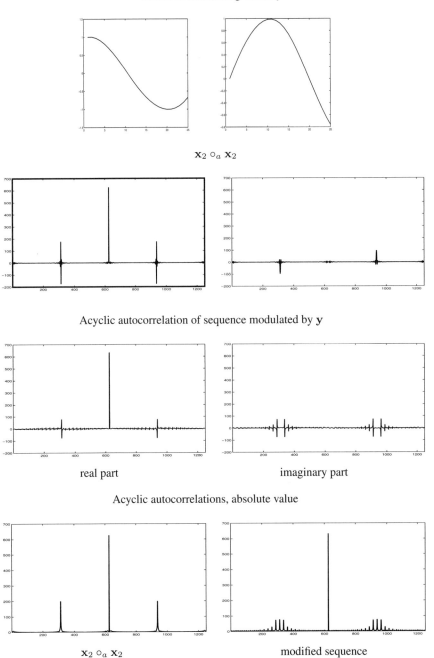

$\mathbf{x}_2 \circ_a \mathbf{x}_2$

Acyclic autocorrelation of sequence modulated by **y**

real part   imaginary part

Acyclic autocorrelations, absolute value

$\mathbf{x}_2 \circ_a \mathbf{x}_2$ modified sequence

Fig. 12.2. Acyclic autocorrelations

12.2 Modulation $N = L^2 R$

Throughout this section $N = LK = L^2 R, R > 1. \Lambda = Perm(L), M = M_{L \times K}$, $Z = Z_{L \times K}$ and $D = D_L, v = e^{2\pi i \frac{1}{L}}$.

We begin by extending the definition of a permutation sequence to a permutation sequence in $\mathbb{C}^N$. Suppose $\phi \in \Lambda$.

Define $\mathbf{e}_\phi \in \mathbb{C}^N$ by

$$Z\mathbf{e}_\phi = [E_\phi \ \cdots \ E_\phi],$$

where E_ϕ is repeated R times. We call $\mathbf{e}_\phi$ a *permutation sequence* in $\mathbb{C}^N$. For $\phi \in \Lambda(L)$, the notation $\mathbf{e}_\phi$ is ambiguous unless there is a reference to the space of $\mathbf{e}_\phi$. As

$$M\mathbf{e}_\phi = \frac{1}{L} \left[F_\phi^* \ \cdots \ F_\phi^* \right],$$

we have $|e_\phi(n)| = \frac{1}{L}, 0 \le n < N$, and

$$\|\mathbf{e}_\phi\|^2 = R.$$

For the remainder of this section, $\mathbf{e}_\phi$ is a permutation sequence in $\mathbb{C}^N$. Set $\mathbf{u} = \mathbf{e}_\phi \circ \mathbf{e}_\phi$.

Theorem 12.5

$$U_0 = R\mathbf{1},$$

$$U_{tL} = tD^{-1}\mathbf{1} + (R - t)\mathbf{1}, \quad 1 \le t < R,$$

and

$$U_k = \mathbf{0}, \quad L \ not \ dividing \ k.$$

Proof Suppose $0 \le m < K$ and write $m = s + rL, 0 \le s < L, 0 \le s < R$. The m-th column of $Z\mathbf{e}_\phi$ is $E_{\phi(s)}$. By the $L \times K$ ZS correlation formula,

$$U_0 = \sum_{s=0}^{L-1} \sum_{r=0}^{R-1} E_{\phi(s)} = R\mathbf{1},$$

and for $1 \le t < R$

$$U_{tL} = D^{-1} \sum_{s=0}^{L-1} \sum_{r=0}^{t-1} E_{\phi(s)} + \sum_{s=0}^{L-1} \sum_{r=t}^{R-1} E_{\phi(s)} = tD^{-1}\mathbf{1} + (R - t)\mathbf{1}.$$

For L not dividing k,

$$U_k = \mathbf{0},$$

completing the proof.

By the theorem, $U_{tL} \ne \mathbf{0}, 1 \le t < R$, and $\mathbf{e}_\phi$ does not satisfy ideal autocorrelation. Modulation is required to *cancel* the columns $U_{tL}, 1 \le t < R$.

For $\mathbf{x} \in \mathbb{C}_1^K$ define $\mathbf{x}_\phi \in \mathbb{C}^N$ by

$$Z\mathbf{x}_\phi = [E_\phi \quad \cdots \quad E_\phi] D(\mathbf{x}).$$

We call $\mathbf{x}_\phi$ a *modulated* permutation sequence in $\mathbb{C}^N$ and say that $\mathbf{x}$ modulates $\mathbf{e}_\phi$. Because $\mathbf{x} \in \mathbb{C}_1^K$,

$$\|\mathbf{x}_\phi\|^2 = R.$$

By Theorem 9.2, the unit discrete chirp in $\mathbb{C}^N$

$$x_u(n) = e^{\pi i u \frac{n^2}{N}} e^{2\pi i f \frac{n}{N}}, \quad n \in \mathbb{Z},$$

is an example of a modulated permutation sequence in $\mathbb{C}^N$, up to a scale factor L.

Suppose $\mathbf{x}_\phi$, $\phi \in \Lambda$, is a modulated permutation sequence in $\mathbb{C}^N$. For $0 \leq s < L$ define the vector $\mathbf{a}(s) \in \mathbb{C}^R$ by

$$a_r(s) = e^{2\pi i \frac{r\phi(s)}{K}} x_{s+rL}, \quad 0 \leq r < R.$$

Notice that extending the expression defining $a_r(s)$ to $0 \leq r < K$ does not result in a vector periodic modulo R, so care must be taken with indexing. As we will see, the transfer from $\mathbf{x}$ to the vectors $\mathbf{a}(s)$, $0 \leq s < L$, is introduced to remove the effect of D^{-1} in the $L \times K$ ZS representation of correlation. Form the autocorrelations

$$\mathbf{c}(s) = \mathbf{a}(s) \circ \mathbf{a}(s), \quad 0 \leq s < L,$$

and set $\mathbf{v} = \mathbf{x}_\phi \circ \mathbf{x}_\phi$.

Theorem 12.6

$$V_0 = R\mathbf{1},$$

$$V_{tL} = \sum_{s=0}^{L-1} e^{-2\pi i \frac{t\phi(s)}{K}} c_t(s) E_{\phi(s)}, \quad 1 \leq t < R,$$

and $V_k = \mathbf{0}$, whenever L does not divide k.

Proof Setting

$$Z\mathbf{x}_\phi = [X_0 \quad \cdots \quad X_{K-1}],$$

we have

$$X_{s+rL} = x_{s+rL} E_{\phi(s)}, \quad 0 \leq s < L,\ 0 \leq r < R.$$

By the ZS correlation formula,

$$V_0 = \sum_{s=0}^{L-1} \sum_{r=0}^{R-1} |x_{s+rL}|^2 E_{\phi(s)} = R \sum_{s=0}^{L-1} E_{\phi(s)} = R\mathbf{1},$$

and $V_k = \mathbf{0}$, whenever L does not divide k.

For $1 \leq t < R$

$$V_{tL} = D^{-1} \sum_{s=0}^{L-1} \sum_{r=0}^{t-1} x_{s+rL} x^*_{s+(R+r-t)L} E_{\phi(s)} + \sum_{s=0}^{L-1} \sum_{r=t}^{R-1} x_{s+rL} x^*_{s+(r-t)L} E_{\phi(s)}.$$

We will rewrite the formula for V_{tL} in terms of the vectors $\mathbf{a}(s)$, $0 \leq s < L$. The transition is based on the following two formulas:

$$v^{-\phi(s)} x_{s+rL} x^*_{s+(R+r-t)L} = e^{-2\pi i \frac{t\phi(s)}{K}} a_r(s) a^*_{R+r-t}(s), \qquad 0 \leq r < t,$$

and

$$x_{s+rL} x^*_{s+(r-t)L} = e^{-2\pi i \frac{t\phi(s)}{K}} a_r(s) a^*_{r-t}(s), \qquad t \leq r < R.$$

Then

$$V_{tL} = \sum_{s=0}^{L-1} e^{-2\pi i \frac{t\phi(s)}{K}} \left(\sum_{r=0}^{t-1} a_r(s) a^*_{R+r-t}(s) + \sum_{r=t}^{R-1} a_r(s) a^*_{r-t}(s) \right) E_{\phi(s)}$$

$$= \sum_{s=0}^{L-1} e^{-2\pi i \frac{t\phi(s)}{K}} c_t(s) E_{\phi(s)},$$

completing the proof.

Corollary 12.1 $\mathbf{x}_\phi$ *satisfies ideal autocorrelation if and only if for $0 \leq s < L$, $\mathbf{a}(s)$ satisfies ideal autocorrelation.*

Proof By Theorem 12.6,

$$F(L)^{-1} V_0 = R\mathbf{e}_0.$$

Suppose for $0 \leq s < L$, $\mathbf{a}(s)$ satisfies ideal autocorrelation. Then

$$c_t(s) = 0, \qquad 1 \leq t < R, \ 0 \leq s < L,$$

and by Theorem 12.6

$$V_{tL} = \mathbf{0}, \qquad 1 \leq t < R,$$

and $V_k = \mathbf{0}$, proving the corollary in one direction.

Conversely, suppose $\mathbf{x}_\phi$ satisfies ideal autocorrelation. Then

$$V_{tL} = 0, \quad 1 \leq t < R.$$

Because the vectors $E_{\phi(s)}, 0 \leq s < L$, are linearly independent, we have

$$c_t(s) = 0, \quad 1 \leq t < R, 0 \leq s < L,$$

completing the proof.

Algorithm 2 *Construct modulated permutation sequences* $\mathbf{x}_\phi$, $\phi \in \Lambda(L)$, *in* $\mathbb{C}^N$ *satisfying ideal autocorrelation.*

- *Choose* $\phi \in \Lambda$ *and construct the permutation sequence* $\mathbf{e}_\phi$ *in* $\mathbb{C}^N$.
- *For* $0 \leq s < L$, *choose a sequence* $\mathbf{a}(s) \in \mathbb{C}^R$ *satisfying ideal autocorrelation.*
- *Define* $\mathbf{x} \in \mathbb{C}^K$ *by*

$$x_{s+rL} = e^{-2\pi i \frac{r\phi(s)}{K}} a_r(s), \qquad 0 \leq r < R, \ 0 \leq s < L.$$

- *Construct* $\mathbf{x}_\phi \in \mathbb{C}^N$ *by*

$$Z\mathbf{x}_\phi = [E_\phi \ \cdots \ E_\phi] D(\mathbf{x}).$$

Example 12.2 For $R = 2$, vectors of the form

$$\begin{bmatrix} e^{2\pi i\theta} \\ ie^{2\pi i\theta} \end{bmatrix},$$

where θ is arbitrary, provide a collection of vectors satisfying ideal autocorrelation. For $L = 5$, choosing $\theta_s = \frac{s}{L}$, $0 \leq s < 5$, we have

$$\mathbf{a}(s) = \begin{bmatrix} w^s \\ iw^s \end{bmatrix}, \qquad w = e^{2\pi i \frac{1}{L}}.$$

Setting $\phi = \phi_0$, Figures 12.3 and 12.4 display the autocorrelation of the resulting sequence. This is compared with the unmodulated sequence. In both cases, the imaginary parts are zero.

By Theorem 6.2 unit discrete chirps of period R can be taken for the sequences $\mathbf{a}(s)$, $0 \leq s < L$, in Algorithm 2. In this way for each permutation sequence $\mathbf{e}_\phi \in \mathbb{C}^N$, $\phi \in \Lambda$, and an integer $R > 1$, we can construct a collection of modulated permutation sequences in $\mathbb{C}^N$ satisfying ideal autocorrelation. The L tuples are not required to consist of distinct unit discrete chirps. Moreover, it can be shown by the ZS characterization of unit discrete chirps in $\mathbb{C}^N$ given in Theorem 9.2 that these collections of modulated permutation sequences in $\mathbb{C}^N$ contain all unit discrete chirps in $\mathbb{C}^N$.

Theorem 12.6 is proved under the assumption that the components of $\mathbf{x}$ have absolute value one. We now relax this condition and extend the class of modulated permutation sequences $\mathbb{C}^N$ as follows. A sequence $\mathbf{x}_\phi \in \mathbb{C}^N$, $\phi \in \Lambda$, is called a *general modulated permutation sequence* in $\mathbb{C}^N$ if

$$Z\mathbf{x}_\phi = [E_\phi \ \cdots \ E_\phi] D(\mathbf{x}), \qquad \mathbf{x} \in \mathbb{C}^K,$$

where E_ϕ is repeated R times and the decimated components of $\mathbf{x}$ have equal norm.

$$\sum_{r=0}^{R-1} |x_{s+rL}|^2 = q, \qquad 0 \leq s < L,$$

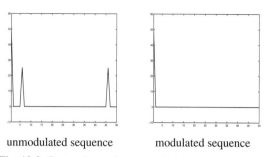

unmodulated sequence   modulated sequence

Fig. 12.3. Comparison of autocorrelations, real parts

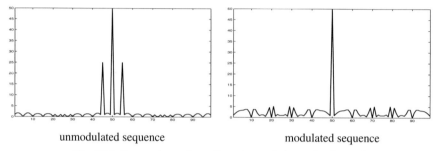

unmodulated sequence   modulated sequence

Fig. 12.4. Comparison of acyclic autocorrelations, absolute value

where q is a constant independent of $0 \le s < L$. Define the vectors $\mathbf{a}(s), 0 \le s < L$, as before. Then

$$\|\mathbf{a}(s)\|^2 = q, \qquad 0 \le s < L.$$

Set $\mathbf{v} = \mathbf{x}_\phi \circ \mathbf{x}_\phi$.

Theorem 12.7 *If* $\mathbf{x}_\phi$, $\phi \in \Lambda$, *is a general modulated permutation sequence in* $\mathbb{C}^N$, *then*

$$V_0 = = q\mathbf{1}$$

$$V_{tL} = \sum_{s=0}^{L-1} e^{-2\pi i \frac{t\phi(s)}{K}} c_t(s) E_{\phi(s)}, \qquad 1 \le t < R,$$

and $V_k = \mathbf{0}$, *otherwise,* $0 \le k < K$.

Proof   The proof is exactly the same as the proof of Theorem 12.6 except we now have

$$V_0 = \sum_{s=0}^{L-1} \sum_{r=0}^{R-1} |x_{s+rL}|^2 E_\phi = q \sum_{s=0}^{L-1} E_\phi = q\mathbf{1}.$$

Arguing as in Corollary 12.1 we have the following result.

Corollary 12.2 *A generalized modulated permutation sequence* $\mathbf{x}_\phi \in \mathbb{C}^N$, $\phi \in \Lambda$, *satisfies ideal autocorrelation if and only if, for* $0 \le s < L$, $\mathbf{a}(s)$ *satisfies ideal autocorrelation.*

Algorithm 3 $\phi \in \Lambda$. *Construct a general modulated permutation sequence* $\mathbf{x}_\phi \in \mathbb{C}^N$ *satisfying ideal autocorrelation.*

• *Choose vectors* $\mathbf{a}(s) \in \mathbb{C}^R$, $0 \le s < L$, *satisfying ideal autocorrelation and*

$$\|\mathbf{a}(s)\|^2 = q, \qquad 0 \le s < L,$$

where q *is independent of* $0 \le s < L$.
• *Construct* $\mathbf{x} \in \mathbb{C}^K$ *by*

$$x_{s+rL} = e^{-2\pi i \frac{r\phi(s)}{K}} a_r(s), \qquad 0 \le s < L, 0 \le r < R.$$

• *Construct* $\mathbf{x}_\phi \in \mathbb{C}^N$ *by*

$$Z\mathbf{x}_\phi = [E_\phi \quad \cdots \quad E_\phi] D(\mathbf{x}).$$

In Chapter 5 we showed that a vector $\mathbf{a} \in \mathbb{C}^R$ satisfies normalized ideal correlation if and only if

$$F\mathbf{a} = \mathbf{u} \in \mathbb{C}_1^R.$$

The components of $\mathbf{a}$ do not necessarily have absolute value one, but $\|\mathbf{a}\| = 1$. To construct a collection of vectors $\mathbf{a}(s)$, $0 \le s < L$, for use in Algorithm 2, we first choose a collection of vectors $\mathbf{u}(s) \in \mathbb{C}_1^R$ and then set

$$\mathbf{a}(s) = F^{-1}\mathbf{u}(s), \qquad 0 \le s < L.$$

By the construction,

$$\|\mathbf{a}(s)\|^2 = 1, \qquad 0 \le s < L.$$

In this way we can parameterize collections of general modulated permutation sequences satisfying ideal autocorrelation.

12.3 Modulation and Ideal Correlation

In this section we establish conditions for ideal correlation of a pair of modulated permutation sequences $\mathbf{x}_\phi$ and $\mathbf{y}_\delta$ in $\mathbb{C}^N$. Pairs of unit discrete chirps are examples. Whenever $*$-permutations are involved, L is an odd integer.

Set $\mathbf{w} = \mathbf{x}_\phi \circ \mathbf{y}_\delta$ and $\gamma = \phi^{-1}\delta$. For $0 \le s < L$, define vectors $\mathbf{a}(s)$ and $\mathbf{b}(s)$ in $\mathbb{C}_1^R$ by

$$a_r(s) = e^{2\pi i \frac{r\phi(s)}{K}} x_{s+rL} \text{ and } b_r(s) = e^{2\pi i \frac{r\delta(s)}{K}} y_{s+rL}, \qquad 0 \le r < R.$$

For $0 \le s, u < L$, set

$$\mathbf{c}(s, u) = \mathbf{a}(s) \circ \mathbf{b}(s - u),$$

where $s - u$ is taken modulo L and

$$\mathbf{c}(s) = \mathbf{c}(s, 0).$$

If γ is a $*$-permutation, we write

$$\phi_{\mathbf{m}} = \left(\phi_0 - \gamma^{-1}\right)^{-1} = (m_0, \quad \ldots, \quad m_{L-1}).$$

Theorem 12.8 *Suppose γ is a $*$-permutation. Then*

$$(\mathbf{x}_\phi, \mathbf{y}_\delta)$$

satisfies ideal correlation if and only if for $0 \le u < L$,

$$(\mathbf{a}(m_u), \mathbf{b}(m_u - u))$$

satisfies ideal correlation, where $m_u - u$ is taken modulo L.

The proof of Theorem 12.8 is carried out in several stages. We begin by making no assumption on $\gamma = \phi^{-1}\delta$.

Theorem 12.9 $W_{tL} = \sum_{s \in \Delta_0(\gamma)} e^{-2\pi i \frac{t\phi(s)}{K}} c_t(s) E_{\phi(s)}, \qquad 0 \le t < R.$

Proof By the $L \times K$ ZS correlation formula

$$W_0 = \sum_{s=0}^{L-1} \sum_{r=0}^{R-1} x_{s+rL} y_{s+rL}^* E_{\phi(s)} E_{\delta(s)} = \sum_{s \in \Delta_0(\gamma)} \sum_{r=0}^{R-1} x_{s+rL} y_{s+rL}^* E_{\phi(s)}.$$

Since $\phi(s) = \delta(s)$, whenever $s \in \Delta_0(\gamma)$, we have

$$x_{s+rL} y_{s+rL}^* = a_r(s) b_r^*(s),$$

and

$$W_0 = \sum_{s \in \Delta_0(\gamma)} \sum_{r=0}^{R-1} a_r(s) b_r^*(s) E_{\phi(s)} = \sum_{s \in \Delta_0(\gamma)} c_0(s) E_{\phi(s)}.$$

For $0 < t < R$

$$W_{tL} = D^{-1} \sum_{s \in \Delta_0(\gamma)} \sum_{r=0}^{t-1} x_{s+rL} y_{s+(R+r-t)L}^* E_{\phi(s)}$$

$$+ \sum_{s \in \Delta_0(\gamma)} \sum_{r=t}^{R-1} x_{s+rL} y_{s+(r-t)L}^* E_{\phi(s)}.$$

We can rewrite the formula for W_{tL} in terms of the vectors $\mathbf{a}(s)$ and $\mathbf{b}(s)$, $0 \le s < L$. As $\phi(s) = \delta(s)$, whenever $s \in \Delta_0(\gamma)$,

$$v^{-\phi(s)}x_{s+rL}y^*_{s+(R+r-t)L} = e^{-2\pi i\frac{t\phi(s)}{K}}a_r(s)b^*_{R+r-t}(s), \quad 0 \le r < t,$$

$$x_{x+rL}y^*_{s+(r-t)L} = e^{-2\pi i\frac{t\phi(s)}{K}}a_r(s)b^*_{r-t}(s), \quad t \le r < R.$$

Placing these formulas into the formula for W_{tL} we have

$$W_{tL} = \sum_{s \in \Delta_0(\gamma)} e^{-2\pi i\frac{t\phi(s)}{K}}\left[\sum_{r=0}^{t-1}a_r(s)b^*_{R+r-t}(s) + \sum_{r=t}^{R-1}a_r(s)b^*_{r-t}(s)\right]E_{\phi(s)},$$

which we can write as

$$W_{tL} = \sum_{s \in \Delta_0(\gamma)} e^{-2\pi i\frac{t\phi(s)}{K}}c_t(s)E_{\phi(s)},$$

completing the proof of the theorem.

Theorem 12.10 *If $tL < u + tL < (t+1)L$, then*

$$W_{t+uL} = \sum_{s \in \Delta^0_u(\gamma)} e^{-2\pi i\frac{(t+1)\phi(s)}{K}}c_{t+1}(s,u)E_{\phi(s)} + \sum_{s \in \Delta^1_u(\gamma)} e^{-2\pi i\frac{t\phi(s)}{K}}c_t(s,u)E_{\phi(s)}.$$

Proof By the $L \times K$ ZS correlation formula, W_{u+tL} is the sum of the four summations

$$D^{-1}\sum_{s=0}^{L-1}\sum_{r=0}^{t-1}x_{s+rL}y^*_{s-u+(R+r-t)L}E_{\phi(s)}E_{\delta(s-u)}$$

$$+ D^{-1}\sum_{s=0}^{u-1}x_{s+tL}y^*_{s-u+RL}E_{\phi(s)}E_{\delta(s-u)}$$

$$+ \sum_{s=u}^{L-1}x_{s+tL}y^*_{s-u}E_{\phi(s)}E_{\delta(s-u)}$$

$$+ \sum_{s=0}^{L-1}\sum_{r=t+1}^{R-1}x_{s+rL}y^*_{s-u+(r-t)L}E_{\phi(s)}E_{\delta(s-u)}.$$

Because

$$E_{\phi(s)}E_{\delta(s-u)} = \begin{cases} E_{\phi(s)} & s \in \Delta_u(\gamma) \\ 0 & \text{otherwise} \end{cases},$$

we can write W_{u+tL} as the sum of six summations

$$D^{-1} \sum_{s \in \Delta_u^0(\gamma)} \sum_{r=0}^{t-1} x_{s+rL} y_{s-u+L+(R+r-t-1)L}^* E_{\phi(s)}$$

$$+ D^{-1} \sum_{s \in \Delta_u^1(\gamma)} \sum_{r=0}^{t-1} x_{s+rL} y_{s-u+(R+r-t)L}^* E_{\phi(s)}$$

$$+ D^{-1} \sum_{s \in \Delta_u^0(\gamma)} x_{s+tL} y_{s-u+L+(R-1)L}^* E_{\phi(s)}$$

$$+ \sum_{s \in \Delta_u^1(\gamma)} x_{s+tL} y_{s-u}^* E_{\phi(s)}$$

$$+ \sum_{s \in \Delta_u^0(\gamma)} \sum_{r=t+1}^{R-1} x_{s+rL} y_{s-u+L+(r-t-1)L}^* E_{\phi(s)}$$

$$+ \sum_{s \in \Delta_u^1(\gamma)} \sum_{r=t+1}^{R-1} x_{s+rL} y_{s-u+(r-t)L}^* E_{\phi(s)}.$$

Combining the 0-th, 2-nd and 4-th summations, and using $\phi(s) = \delta(s - u)$, whenever s is in $\Delta_u(\gamma)$, we have, after simplifying,

$$\sum_{s \in \Delta_u^0(\gamma)} e^{-2\pi i \frac{(t+1)\phi(s)}{K}} c_{t+1}(s, u) E_{\phi(s)}.$$

Combining the rest, we have

$$\sum_{s \in \Delta_u^1(\gamma)} e^{-2\pi i \frac{t\phi(s)}{K}} c_t(s, u) E_{\phi(s)},$$

completing the proof of the theorem.

If γ is a $*$-permutation and $\phi_{\mathbf{m}} = (\phi_0 - \gamma^{-1})^{-1}$, we have $\Delta_u(\gamma) = \{m_u\}$, $0 \le u < L$. By Theorems 12.1, 12.2 and 12.10 we have the following result.

Theorem 12.11 *If γ is a $*$-permutation, then for $0 \le t < R$,*

$$W_{tL} = e^{-2\pi i \frac{t\phi(m_0)}{K}} c_t(m_0) E_{\phi(m_0)}.$$

For $tL < u + tL < (t+1)L$

$$W_{u+tL} = e^{-2\pi i \frac{(t+1)\phi(m_u)}{K}} c_{t+1}(m_u, u) E_{\phi(m_u)}, \qquad m_u < u,$$

and

$$W_{u+tL} = e^{-2\pi i \frac{t\phi(m_u)}{K}} c_t(m_u, u) E_{\phi(m_u)}, \qquad u \le m_u.$$

To complete the proof of Theorem 12.8, note that by Theorem 12.11

$$|c_t(m_u, u)| = \sqrt{R}, \qquad 0 \le t < R,\ 0 \le s,\ u < L,$$

if and only if

$$|w(n)| = \frac{\sqrt{R}}{L}, \qquad 0 \le n < N.$$

Algorithm 4 $\phi, \delta \in \Lambda$ such that $\gamma = \phi^{-1}\delta$ is a $*$-permutation. Construct modulated permutation sequences $\mathbf{x}_\phi$ and $\mathbf{y}_\delta$ in $\mathbb{C}^N$ such that $(\mathbf{x}_\phi, \mathbf{y}_\delta)$ satisfies ideal correlation.

- For $0 \le s < L$, construct vectors $\mathbf{a}(s)$ and $\mathbf{b}(s)$ in $\mathbb{C}_1^R$ such that for $0 \le u < L$

$$(\mathbf{a}\,(m_u)\,, \mathbf{b}\,(m_u - u))$$

  satisfies the ideal correlation property, where $m_u - u$ is taken modulo L.
- Construct vectors $\mathbf{x}$ and $\mathbf{y}$ in $\mathbb{C}_1^K$ by

$$x_{s+rL} = e^{-2\pi i \frac{r\phi(s)}{K}} a_r(s),$$

and

$$y_{s+rL} = e^{-2\pi i \frac{r\delta(s)}{K}} b_r(s), \qquad 0 \le s < L,\ 0 \le r < R.$$

- Construct $\mathbf{x}_\phi$ and $\mathbf{y}_\delta$ in $\mathbb{C}^N$ by

$$Z\mathbf{x}_\phi = (Z\mathbf{e}_\phi)\,D(\mathbf{x}) \quad \text{and} \quad Z\mathbf{y}_\delta = (Z\mathbf{e}_\delta)\,D(\mathbf{y}).$$

Example 12.3 $L = 5$ and $\gamma = \mu_3$. Then

$$\phi_\mathbf{m} = \left(\phi_0 - \gamma^{-1}\right)^{-1} = (0\ 4\ 3\ 2\ 1).$$

The vectors in $\mathbb{C}^R$

$$\mathbf{a}(s) \quad \text{and} \quad \mathbf{b}(s), \qquad 0 \le s < 5,$$

satisfy the conditions in Theorem 12.8 if

$$(\mathbf{a}(0), \mathbf{b}(0)),\ \ (\mathbf{a}(4), \mathbf{b}(3)),\ \ (\mathbf{a}(3), \mathbf{b}(1)),\ \ (\mathbf{a}(2), \mathbf{b}(4)),\ \ (\mathbf{a}(1), \mathbf{b}(2))$$

satisfy ideal correlation.

By Theorem 6.3 pairs of unit discrete chirps

$$(\mathbf{x}_u, \mathbf{x}_v), \qquad u - v \in U_R,$$

can be taken for the sequence pairs in Algorithm 4. In this way for each permutation sequence pair in $\mathbb{C}^N$

$$(\mathbf{e}_\phi, \mathbf{e}_\delta), \qquad \gamma = \phi^{-1}\delta \in \Lambda^*$$

and an integer $R > 1$, we can construct a collection of modulated permutation sequence pairs in $\mathbb{C}^N$ satisfying ideal correlation parameterized by L tuples of unit discrete chirp pairs in $\mathbb{C}^R$ satisfying ideal correlation. The L tuples are not required to consist of distinct unit discrete chirp pairs.

Example 12.4 $L = 5$ and $\gamma = \mu_3$. Since γ is a $*$-permutation, we can choose $\phi = \phi_0$ and $\delta = \gamma$. For $R = 3$, we have the pair of discrete chirps satisfying ideal correlation, $\mathbf{x}_1$ and $\mathbf{x}_2$.

$$\mathbf{x}_1 = \begin{bmatrix} 1 \\ w \\ 1 \end{bmatrix}, \qquad \mathbf{x}_2 = \begin{bmatrix} 1 \\ w \\ w \end{bmatrix}, \qquad w = e^{2\pi i \frac{1}{3}}.$$

The discrete carrier frequency for $\mathbf{x}_1$ is $\frac{1}{2}$ and that for $\mathbf{x}_2$ is 0. Set

$$\mathbf{a}(s) = \mathbf{x}_1, \qquad \mathbf{b}(s) = \mathbf{x}_2, \qquad 0 \le s < L.$$

Figures 12.5 and 12.6 display the resulting pair of modulated permutation sequences. They satisfy ideal correlation. Figure 12.7 displays the acyclic correlations. Only the real parts are displayed, as the sidelobes are a good indication of the range of values of the imaginary parts.

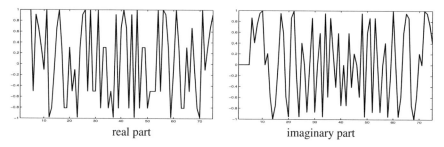

real part   imaginary part

Fig. 12.5. Modulated permutation sequences

Theorem 12.8 along with Algorithm 3 can be extended to the class of general modulated permutation sequences. The derivation of the cross correlation results are not affected by this extension. In Algorithm 3 we can now choose $\mathbf{a}(s)$ and $\mathbf{b}(s)$, $0 \le s < L$, such that the norms

$$\|\mathbf{a}(s)\|^2 \quad \text{and} \quad \|\mathbf{b}(s)\|^2, \qquad 0 \le s < L,$$

are independent of $0 \le s < L$.

Algorithm 1 in Chapter 5 constructs normalized ideal correlation pairs

$$(\mathbf{a}, \mathbf{b}), \qquad \mathbf{a}, \mathbf{b} \in \mathbb{C}^R$$

by the following procedure. Choose $\mathbf{w} \in \mathbb{C}_1^R$ such that

$$\frac{1}{\sqrt{R}} F(R) \mathbf{w} \in \mathbb{C}_1^R.$$

Each factorization

$$\mathbf{w} = \mathbf{u} \mathbf{v}^*, \qquad \mathbf{u}, \mathbf{v} \in \mathbb{C}_1^R,$$

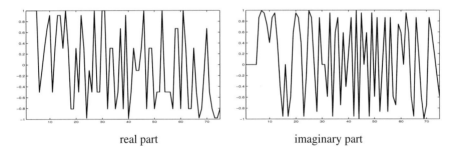

real part   imaginary part

Fig. 12.6. Modulated permutation sequences

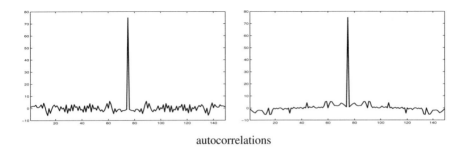

autocorrelations

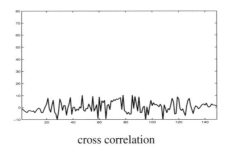

cross correlation

Fig. 12.7. Acyclic correlations of modulated permutation sequences, $N = 75$

leads to a normalized ideal correlation pair by setting

$$F(R)\mathbf{a} = \mathbf{u} \quad \text{and} \quad F(R)\mathbf{b} = \mathbf{v}.$$

Given such a $\mathbf{w}$ (or perhaps several) we can construct a collection of normalized ideal correlation pairs by first choosing a collection of factorizations

$$\mathbf{w} = \mathbf{u}(s)\mathbf{v}(s)^*, \quad 0 \le s < L, \ \mathbf{u}(s), \ \mathbf{v}(s) \in \mathbb{C}_1^R,$$

and then setting

$$F(R)\mathbf{a}(s) = \mathbf{u}(s) \text{ and } F(R)\mathbf{b}(s) = \mathbf{v}(s), \qquad 0 \le s < L.$$

For the algorithm the vectors in the pairs may have to be permuted depending upon γ.

13

Sequence Sets

$L > 1$ is an odd integer. $\Lambda = Perm(L)$. In this chapter we develop design strategies for constructing signal sets of modulated permutation sequences, including discrete chirp sequences, of period L^2 satisfying pairwise ideal correlation. As modulation, in this case, does not affect correlation properties, we restrict the discussion to signal sets of permutation sequences. The extension to sequences of period RL^2 can be carried out by the methods of Section 12.2.

The methods presented in this chapter construct a very large collection of sequence sets. While a particular application may require only one collection of a few sequences, the flexibility of choosing a collection to best fit the application is significantly increased.

13.1 Ideal Correlation Roots

A set of $*$-permutations
$$S = \{\phi_j : 1 \leq j < J\}$$
is called an *ideal correlation root* (ICR) of *size* $J - 1$ if
$$\phi_j(0) = 0, \quad 1 \leq j < J,$$
and
$$\phi_j^{-1}\phi_k \in \Lambda^*, \quad 1 \leq j < k < J.$$
S is called a *standard* ICR (SICR) if
$$\phi_j = \mu_{u_j}, \quad 1 \leq j < J.$$
In the case of an SICR,
$$u_j, \; 1 - u_j \in U_L, \quad 1 \leq j < J,$$
and

M. An et al., *Ideal Sequence Design in Time-Frequency Space*,
DOI 10.1007/978-0-8176-4738-4_13,
© Birkhäuser Boston, a part of Springer Science+Business Media, LLC 2009

$$u_j - u_k \in U_L, \qquad 1 \leq j < k < J.$$

We develop the theory for general ICRs and use SICRs in examples. In the next section, we compare the number of signal sets constructed from ICRs with signal sets constructed from SICRs. As the number of $*$-permutations can be substantially greater than the number of unit $*$-permutations of the same periodicity, the number of signal sets constructed from ICRs is substantially greater than the number of signal sets constructed from SICRs. This advantage seems to increase as the periodicity L^2 increases and as the number of prime factors of L increases. However, both theory and numerics are far from being understood.

Unless otherwise specified,

$$S = \{\phi_j : 1 \leq j < J\}$$

is an arbitrary ICR of size $J - 1$. Each $\phi \in \Lambda$ determines a *branch* S_ϕ of *size* J over S defined by

$$S_\phi = \{\phi\phi_j : 0 \leq j < J\}, \qquad \phi_0 \text{ identity permutation.}$$

By Theorem 11.8 a branch S_ϕ over S defines a signal set of permutation sequences of size J satisfying pairwise ideal correlation. Each permutation in S fixes 0, but as this is not necessarily the case for ϕ, the permutations in S_ϕ need not fix 0. An extension to roots not necessarily fixing 0 will be discussed at the end of the section.

Example 13.1 For $S = \{\phi_1, \phi_2\}$,

$$S_\phi = \{\phi, \phi\phi_1, \phi\phi_2\}.$$

Note that for $\phi, \phi' \in \Lambda$

$$\phi S_{\phi_0} = S_\phi,$$

and

$$(\phi\phi') S_{\phi_0} = \phi S_{\phi'} = S_{\phi\phi'}.$$

Distinct ϕ and ϕ' in Λ can define identical branches over S,

$$S_\phi = S_{\phi'}.$$

In this case $S_{\phi^{-1}\phi'} = S_{\phi_0}$. Define

$$G(S) = \{\phi \in \Lambda \ : \ S_\phi = S_{\phi_0}\}.$$

Because $\phi_0 \in G(S)$ and

$$\phi, \phi' \in G(S) \text{ implies } \phi\phi' \in G(S),$$

then $G(S)$ is a subgroup of Λ.

Suppose $\phi \in G(S)$. Because $\phi_0 \in S_\phi$, we have

$$\phi = \phi_j^{-1}$$

for some $0 \leq j < J$ and

$$G(S) = \{\phi_j \ : \ \phi_j S_{\phi_0} = S_{\phi_0}, \quad 0 \leq j < J\}.$$

In particular,

$$|G(S)| \leq J.$$

Example 13.2 $L = 5$. $S = \{\mu_2, \ \mu_3\}$ is an SICR and $S_{\phi_0} = \{\mu_1, \ \mu_2, \ \mu_3\}$. Because

$$\mu_2 S_{\phi_0} = \{\mu_2, \ \mu_4, \ \mu_1\} \neq S_{\phi_0}$$

and

$$\mu_3 S_{\phi_0} = \{\mu_3, \ \mu_1, \ \mu_4\} \neq S_{\phi_0},$$

we have $G(S) = \{\phi_0\}$.

Example 13.3 $L = 11$. $S = \{\mu_2, \ \mu_9, \ \mu_{10}\}$ is an SICR. Because

$$\mu_2 S_{\phi_0} = \{\mu_2, \ \mu_4, \ \mu_7, \ \mu_9\} \neq S_{\phi_0},$$

$$\mu_9 S_{\phi_0} = \{\mu_9, \ \mu_7, \ \mu_4, \ \mu_2\} \neq S_{\phi_0},$$

but

$$\mu_{10} S_{\phi_0} = \{\mu_{10}, \ \mu_9, \ \mu_2, \ \mu_1\} \neq S_{\phi_0},$$

we have $G(S) = \{\phi_0, \ \mu_{10}\}$.

In general if S_{ϕ_0} is a group of permutations, then

$$G(S) = S_{\phi_0}.$$

Example 13.4 $L = 7$. $S = \{\mu_u : 2 \leq u < G\}$ is an SICR and

$$S_{\phi_0} = \{\mu_u \ : \ u \in U_7\}$$

is a group. Then

$$G(S) = \{\mu_u : u \in U_7\}.$$

Example 13.5 $L = 11$. $S = \{\mu_3, \ \mu_9, \ \mu_5, \ \mu_4\}$ is an SICR and

$$S_{\phi_0} = \{\mu_1, \ \mu_3, \ \mu_9, \ \mu_5, \ \mu_4\}$$

is a group. Then

$$G(S) = \{\mu_1, \ \mu_3, \ \mu_9, \ \mu_5, \ \mu_4\}.$$

Denote by

$$\Lambda/G(S)$$

the collection of all subsets of Λ of the form

$$\phi G(S) = \{\phi\phi' : \phi' \in G(S)\}.$$

$\Lambda/G(S)$ is called the collection of *right cosets* of $G(S)$ in Λ. We can also call $\phi G(S)$ the $G(S)$-orbit through ϕ. Suppose

$$\phi G(S) = \phi' G(S), \quad \phi, \; \phi', \in \Lambda.$$

Because $\phi_0 \in G(S)$, we have $\phi \in \phi' G(S)$ and $(\phi')^{-1}\phi \in G(S)$, implying

$$(\phi')^{-1}\phi \in G(S) \quad \text{and} \quad \phi^{-1}\phi' \in G(S).$$

The converse also holds by reversing the steps and we have the important result

$$\phi G(S) = \phi' G(S)$$

if and only if

$$(\phi')^{-1}\phi \in G(S) \quad \text{and} \quad \phi^{-1}\phi' \in G(S).$$

The construction of collections of right cosets and left cosets is a central tool in group theory. It appears implicitly or explicitly in many application fields including digital signal processing and error-correcting coding, usually for commutative groups, where the constructions lead to a commutative group. For instance, the most known example is the group $\mathbb{Z}/N$ of integers modulo N. We require the following result. The collection of right cosets $\Lambda/G(S)$ forms a partition of Λ and the number of subsets in the collection is

$$\frac{|\Lambda|}{|G(S)|} = \frac{L!}{|G(S)|},$$

which is bounded below by $\frac{L!}{J}$.

Because $S_\phi = S_{\phi'}$ if and only if $\phi^{-1}\phi' \in G(S)$ if and only if

$$\phi G(S) = \phi' G(S),$$

the mapping

$$\phi G(S) \to S_\phi$$

is a well-defined one-to-one mapping of $\Lambda/G(S)$ onto the collection of distinct branches over S. In particular, the number of branches over S equals $\frac{L!}{G(S)}$ and is bounded below by $\frac{L!}{J}$. The lower bound is achieved when S_{ϕ_0} is a group of permutations.

Example 13.6 $L = 5$. The number of branches over

$$S = \{\mu_2, \; \mu_3\}$$

is $5!$.

Example 13.7 $L = 11$. The number of branches over

$$S = \{\mu_2, \; \mu_4, \; \mu_{10}\}$$

is $\frac{11!}{2}$.

Example 13.8 $L = 7$. The number of branches over

$$S = \{\mu_u : 2 \le u < 6\}$$

is $\frac{7!}{6}$.

Example 13.9 $L = 11$. The number of branches over

$$S = \{\mu_3,\ \mu_9,\ \mu_5,\ \mu_4\}$$

is $\frac{11!}{5}$.

Distinct roots can have a common branch. Suppose

$$S' = \{\phi'_1,\ \ldots,\ \phi'_{J-1}\}$$

is a second ICRs of size $J - 1$. If S and S' have a common branch

$$S_\phi = S'_{\phi'}, \quad \text{some } \phi,\ \phi' \in \Lambda,$$

then

$$S_{\phi_0} = S'_{\phi^{-1}\phi'}$$

and the branches of S and S' are identical.

Two ICRs are said to be *equivalent* if their branches are identical. This is an equivalence relation on the collection of ICRs of size $J - 1$. Equivalent ICRs provide different parameterizations for their collection of branches and to the corresponding collection of signal sets.

We give a procedure for describing the collection of all ICRs equivalent to S. An ICR S' of size $J - 1$ is equivalent to S if and only if there exists $\phi \in \Lambda$ such that

$$S_{\phi_0} = S'_\phi.$$

We begin by describing a procedure for constructing such an S' and then show that the procedure describes all ICRs equivalent to S.

Define

$$S' = \{\phi_1^{-1},\ \phi_1^{-1}\phi_2,\ \ldots,\ \phi_1^{-1}\phi_{J-1}\}.$$

Then S' is an ICR and

$$S_{\phi_0} = S'_{\phi_1},$$

where S' is equivalent to S.

Up to reindexing, we show that this construction leads to the collection of all ICRs equivalent to S.

Suppose

$$S' = \{\phi'_1,\ \ldots,\ \phi'_{J-1}\}$$

is an ICR equivalent to S. Then

$$S_{\phi_0} = S'_\phi, \quad \text{some } \phi \in \Lambda.$$

If $\phi = \phi_0$, then $S' = S$. Otherwise, $\phi \in S_{\phi_0}$, and we can reindex to obtain

$$\phi = \phi_1.$$

Then

$$S'_\phi = \{\phi_1, \ \phi_1\phi'_1, \ \ldots, \ \phi_1\phi'_{J-1}\}.$$

As $\phi_0 \in S'_\phi$, we can reindex to have

$$\phi_0 = \phi_1\phi'_1.$$

Then

$$\{\phi_2, \ \ldots, \ \phi_{J-1}\} = \{\phi_1\phi'_2, \ \ldots, \ \phi_1\phi'_{J-2}\}.$$

Reindexing we have

$$\phi'_j = \phi_1^{-1}\phi_j, \quad 2 \leq j < J,$$

and we have shown that, up to indexing,

$$S' = \{\phi_1^{-1}, \ \phi_1^{-1}\phi_2, \ \ldots, \ \phi_1^{-1}\phi_{J-1}\}$$

as claimed.

A problem remains. The preceding construction describes the equivalence class containing S, but the J ICRs constructed need not be distinct. We do know that if $CL(S)$ is the equivalence class containing S, then

$$|CL(S)| \leq J.$$

We study the problem in more detail for SICRs. Suppose S is an SICR defined by the elements

$$u_1, \ \ldots, \ u_{J-1}.$$

The equivalence class of SICRs containing S is computed, up to indexing, by repeating the mapping

$$\{u_1, \ \ldots, \ u_{J-1}\} \to \{u_1^{-1}, \ u_1^{-1}u_2, \ \ldots u_1^{-1}u_{J-1}\}.$$

Example 13.10 $L = 5$. The SICRs of size 2

$$\{\mu_2, \ \mu_3\}, \ \{\mu_3, \ \mu_4\}, \ \{\mu_2, \ \mu_4\}$$

are equivalent and form an equivalence class.

Example 13.11 $L = 7$. The following collections:

$$\{\{\mu_2, \ \mu_3\}, \ \{\mu_4, \ \mu_5\}, \ \{\mu_3, \ \mu_5\}\},$$

$$\{\{\mu_2, \ \mu_5\}, \ \{\mu_3, \ \mu_6\}, \ \{\mu_4, \ \mu_6\}\},$$

and

$$\{\{\mu_2, \ \mu_6\}, \ \{\mu_3, \ \mu_4\}, \ \{\mu_3, \ \mu_6\}\}$$

are equivalence classes. The single element collection

$$\{\{\mu_2, \ \mu_4\}\}$$

is an equivalence class.

In general the SICR

$$\{\mu_{v_1}, \, \mu_{v_2}\}$$

forms an equivalence class of order 1 if and only if

$$v_1^3 = 1 \ \text{ and } \ v_2 = v_1^2.$$

For $L = p$, p an odd prime such that $p \neq 1 \bmod 3$, there exists no SICR of size 2 forming an equivalence class of order 1. For $p \equiv 1 \bmod 3$, there exists a unique SICR of size 2 forming an equivalence class of order 1.

Example 13.12 $L = 13$. The SICR

$$\{\mu_3, \, \mu_9\}$$

forms an equivalence class of order 1.

Example 13.13 $L = 11$. The SICR of size 3

$$\{\mu_2, \, \mu_9, \, \mu_{10}\}$$

forms an equivalence class of order 1.

In general, the SICR of size 3

$$\{\mu_r, \, \mu_s, \, \mu_r s\}, \qquad r^2 = 1,$$

forms an equivalence class of order 1.

Example 13.14 $L = 13$. The SICR of size 3

$$\{\mu_2, \, \mu_{11}, \, \mu_{12}\}$$

forms an equivalence class of order 1. The SICR of size 3

$$\{\mu_5, \, \mu_8, \, \mu_{12}\}$$

forms an equivalence class of order 1.

In general, the SICR of size 3

$$\{\mu_r, \, \mu_{r^2}, \, \mu_{r^3}\}, \qquad r^2 \neq 1 \ \text{ and } r^4 = 1,$$

forms an equivalence class of order 1.

The definition of an ICR can be extended to include shifts. For

$$\mathbf{m} = (m_0, \, m_1, \, \ldots, \, m_{J-1}) \in (\mathbb{Z}/L)^J,$$

define the *shift* ICR $S(\mathbf{m})$ over S by

$$S(\mathbf{m}) = \{\phi_j \sigma^{m_j} : 1 \leq j < J\}, \quad \sigma \text{ the shift permutation.}$$

By Theorem 10.3,
$$\phi_j \sigma^{m_j} \in \Lambda^*, \qquad 1 \le j < J,$$
and
$$\sigma^{-m_j} \phi_j^{-1} \phi \quad \text{or} \quad \sigma^{m_k} \in \Lambda^*, \qquad 1 \le j < k < J.$$
This implies that the *branch* $S_\phi(\mathbf{m})$ over $S(\mathbf{m})$ defined by
$$S_\phi(\mathbf{m}) = \{\phi \phi_j \sigma^{m_j} : 0 \le j < J\}$$
determines a signal set of permutation sequences
$$\{\mathbf{e}'_\phi : \phi' \in S_\phi(\mathbf{m})\}$$
satisfying pairwise ideal correlation.

The shift ideal correlation roots over S are distinct. Suppose
$$S(\mathbf{m}) = S(\mathbf{n}), \qquad \mathbf{m}, \ \mathbf{n} \in (\mathbb{Z}/L)^J.$$
Then for some $0 \le j, \ k < J$,
$$\phi_j \sigma^{m_j} = \phi_k \sigma^{m_k},$$
and
$$\phi_k^{-1} \phi_j = \sigma^{n_k - m_j}.$$
Because $\phi_k^{-1} \phi_j(0) = 0$,
$$\phi_k^{-1} \phi_j = \sigma^{m_k - n} = \phi_0,$$
proving the claim. As a result, there are L^{J-1} distinct shift ideal correlation roots over S, defining L^J signal sets of permutation sequences satisfying pairwise ideal correlation.

13.2 Numerics

Unless otherwise specified, permutations are in $\Lambda = Perm(L)$. In this section we compare the number of ICRs with the number of SICRs for several periodicities L^2 and size $J - 1$. The study is by no means complete either theoretically or numerically. The main problem is that we have no constructive method for computing the $*$-permutations and their number grows very quickly with L.

The following result gives an upper bound on the number of elements in an ICR.

Theorem 13.1 *If there exists an ICR of size $J - 1$, then $J < L$.*

Proof Suppose $\phi_1, \ldots, \phi_{J-1}$ is an ICR of size $J - 1$. Set
$$a_j = \phi_j(1), \qquad 1 \le j < J.$$
Because $\phi_j(0) = 0$ and $\phi_j \in \Lambda^*$, we have

$$a_j \in \{2, \ldots, L-1\}, \qquad 1 \le j < J.$$

If $a_j = a_k$, then

$$\phi_j^{-1}\phi_k(1) = 1.$$

As $\phi_j^{-1}\phi_k$ is a $*$-permutation, we have $j = k$, showing that the numbers

$$a_1, \ldots, a_{J-1}$$

are distinct, proving the theorem.

We begin with the unit $*$-permutations and the SICRs for the cases $L = p$ and $L = p^2$, p an odd prime.

For $L = p$, p an odd prime, the set of unit $*$-permutations

$$\Lambda_0^* = \{\mu_u : 2 \le u < p\}$$

has order $p - 2$. The number of SICRs of size $J - 1$ is

$$\frac{(p-2)(p-3)\cdots(p-J)}{(J-1)!}, \qquad 2 \le J < p.$$

By Theorem 10.2, there is a unique maximum size SICR of size $J - 1 = p - 2$.

$$\{\mu_u : 2 \le u < p\}.$$

Example 13.15 $L = 5$.

$$\Lambda_0^* = \{\mu_2, \mu_3, \mu_4\}.$$

The SICRs of size $J - 1 = 2$ are

$$\{\mu_2, \mu_3\}, \quad \{\mu_2, \mu_4\}, \quad \{\mu_3, \mu_4\}.$$

The unique maximum size SICR is

$$\{\mu_2, \mu_3, \mu_4\}.$$

For $L = p^2$, p an odd prime, the set of unit $*$-permutations

$$\Lambda_0^* = \{\mu_u : 0 \le u < p^2, \ u, \ 1 - u \in U_p\}$$

has order $(p-2)p$. The number of SICRs of size $J - 1$ is

$$\frac{(p-2)(p-3)\cdots(p-J)}{(J-1)!} p(p-1)\cdots(p-J+2), \qquad 2 \le J < p.$$

The maximum size SICR is

$$J - 1 = p - 2$$

and the number of SICRs achieving this maximum is

$$\frac{p!}{2!}.$$

Example 13.16 $L = 9$.

$$\Lambda_0^*(9) = \{\mu_2, \ \mu_5, \ \mu_8\}.$$

The maximum size SICR is one and there are three SICRs achieving this maximum,

$$\{\mu_2\}, \ \ \{\mu_5\}, \ \ \{\mu_8\}.$$

Example 13.17 $L = 25$. The maximum size SICR is 3 and there are 60 SICRs achieving this maximum.

Another interesting example is $L = 15$. The maximum size SICR is 1 and there are 3 SICRs achieving this maximum. Tables 13.1 and 13.2 list the number of SICRs as a function of J for $L = 11$ and $L = 25$.

Table 13.1. Number of SICRs of size $J - 1$, $L = 11$

J	2	3	4	5	6	7	8	9	10
	9	36	84	126	126	84	36	9	1

Table 13.2. Number of SICRs of size $J - 1$, $L = 25$

J	2	3	4
	15	60	60

Table 13.3 compares the number of unit $*$-permutations with the number of $*$-permutations normalized by 0 mapping to 0.

Table 13.3. Number of SICRs and ICRs of size $J - 1$

L	5	7	9	11	13	15
SICR	3	5	3	9	11	3
ICR	3	19	2025	3441	79,259	1,026,283

For $L = 7$, the 14 nonunit $*$-permutations mapping 0 to 0 are listed in Table 13.4.

Table 13.4. Nonunit $*$-permutations $L = 7$

0	0	0	0	0	0	0	0	0	0	0	0	0	0
2	2	3	3	3	3	4	4	5	5	5	5	6	6
5	6	1	5	6	6	3	6	1	3	3	4	3	4
1	5	6	2	2	4	1	2	4	2	6	1	5	2
6	3	5	1	5	2	6	5	6	6	2	3	1	5
4	1	2	6	1	1	2	3	3	1	4	6	4	1
3	4	4	4	4	5	5	1	2	4	1	2	2	3

Although the numerics are by no means complete, there are certain trends that can be inferred by the tables. It seems that as L increases and/or becomes more composite the ratio

$$\frac{\text{number of }*\text{-permutations } (\pi(0) = 0)}{\text{number of standard }*\text{-permutations}}, \quad (J = 2)$$

increases as does the ratio

$$\frac{\text{number of ICRs of size } J - 1}{\text{number of SICRs of size } J - 1}$$

for J below some bound. After this bound the ratio is one. This bound appears to increase by more than first order with the increase in L.

Table 13.5 compares the number of ICRs and SICRs for different L and J. Note that in the case $L = 15$ the maximum size SICR is 1, but there exist many ICRs of size 2.

Table 13.5. Number of SICRs and ICRs achieving order $J - 1$

L	$J - 1$	number of sets in Λ_0^*	in Λ^*
5	1	3	3
5	2	2	2
5	3	1	1
7	1	5	19
7	2	10	10
7	3	10	10
7	4	5	5
7	5	1	1
9	1	3	2025
11	1	9	3441
11	2	36	2016
11	3	168	788
11	4	126	126
11	5	126	126
11	6	84	84
11	7	36	36
11	8	9	9
11	9	1	1
11	9	1	1
13	1	11	$79,259$
13	2	55	$395,242$
13	3	165	$7,250$
13	4	330	$3,060$
13	5	462	$1,010$
13	6	462	462
13	7	330	330
13	8	165	165
13	9	55	55
13	10	11	11
15	1	3	$1,026,283$
15	2	0	$> 4,000,000$

Echo Analysis

We assume the following simple monostatic radar and target environment model. An analog pulse of time duration T is transmitted. The target environment consists of stationary point targets, perhaps with varying scattering amplitudes, whose travel times from the transmitter are integer multiples of $\frac{T}{2N}$. Sampling the echoes at points $\frac{nT}{N}$, $n \in \mathbb{Z}$, produces a sequence which is a linear combinations of shifts of the sampled transmitted signal. The scattering amplitudes are the coefficients of the linear combinations.

We work solely with sequences assuming these sequences can be formed by sampling an analog envelope.

In this chapter we address the problem of spatial localization of multiple targets from echoes of modulated permutation sequences including discrete chirps. Zak space, which played a critical role in sequence design, plays a new role as an image space for echo analysis. The echoes resulting from the reflections of a discrete chirp pulse on a collection of stationary point targets will be a set of parallel lines corrupted by noise. The location and number of lines in Zak space indicate the number and radial distances of the targets, irrespective of the values on the lines. In the presence of severe noise these lines can be obstructed or broken, and may no longer be collinear. Image reconstruction can be applied to find the best fit to these corrupted lines in the Zak space (ZS) image. In [39] the methods developed in this chapter were applied to the problem of identifying dielectric materials from chirp echoes.

We begin with a review of the cyclic case. For applications it is necessary to extend results on cyclic shifts to linear or aperiodic shifts. We do this by zero-padding.

Suppose $\mathbf{x} \in \mathbb{C}^N$ and $\mathbf{x}^J$ is the zero-padding of $\mathbf{x}$ to $\mathbb{C}^{NJ}$,

$$\mathbf{x}^J = \mathbf{e}_0 \otimes \mathbf{x}, \qquad \mathbf{e}_0 = \mathbf{e}_0^J.$$

Viewing $\mathbf{x}^J$ as a sequence whose values outside the indices $0 \leq n < N$ vanish, we can identify the cyclic shifts

$$S_{NJ}^n \mathbf{x}^J, \qquad 0 \leq n < (J-1)N,$$

with the corresponding linear shifts on the sequence corresponding to $\mathbf{x}^J$.

M. An et al., *Ideal Sequence Design in Time-Frequency Space*,
DOI 10.1007/978-0-8176-4738-4_14,
© Birkhäuser Boston, a part of Springer Science+Business Media, LLC 2009

Example 14.1 $N = 2$ and $J = 3$. Then

$$\left(\mathbf{x}^3\right)^T = [x_0 \ x_1 \ 0 \ 0 \ 0 \ 0],$$

$$\left(S_6\mathbf{x}^3\right)^T = [0 \ x_0 \ x_1 \ 0 \ 0 \ 0],$$

$$\left(S_6^2\mathbf{x}^3\right)^T = [0 \ 0 \ x_0 \ x_1 \ 0 \ 0],$$

$$\left(S_6^3\mathbf{x}^3\right)^T = [0 \ 0 \ 0 \ x_0 \ x_1 \ 0].$$

14.1 Cyclic Shifts

$N = LK = L^2J$, where $L > 1$, $J \geq 1$ are integers. $S = S_L$, $Z = Z_L$ and $\Lambda = Perm(L)$.

Suppose $\mathbf{u} \in \mathbb{C}^N$. A *cyclic echo* of $\mathbf{u}$ is a linear combination,

$$\mathbf{v} = \sum_{n=0}^{N-1} a_n S_N^n \mathbf{u}.$$

The problem of echo analysis is to determine the coefficients of the expansion given an echo $\mathbf{v}$.

If $\mathbf{u}$ satisfies ideal autocorrelation, then the collection of cyclic shifts of $\mathbf{u}$ is orthogonal, and in the absence of noise the coefficients can be computed directly by the cross correlation or inner products ($\|u\| = 1$),

$$a_m = \mathbf{v} \circ \mathbf{u}(m) = \langle \mathbf{v}, S_N^m \mathbf{u} \rangle, \qquad 0 \leq m < N.$$

Suppose X is an $L \times K$ matrix. The *support* of X denoted by $supp \ X$ is defined by

$$supp \ X = \{(l, k) \in \mathbb{Z}/L \times \mathbb{Z}/K : X(l, k) \neq 0\}.$$

Example 14.2 If

$$\phi = (0 \ 2 \ 4 \ 1 \ 3),$$

then

$$supp \ E_\phi = \{(0,0), (1,3), (2,1), (3,4), (4,2)\}.$$

Suppose $\mathbf{x}_\phi$, $\phi \in \Lambda$, is a modulated permutation sequence in $\mathbb{C}^N$. Ideal autocorrelation is not assumed.

$$Z\mathbf{x}_\phi = [E_\phi \ \cdots \ E_\phi] D(\mathbf{c}), \qquad \mathbf{c} \in \mathbb{C}_1^K.$$

The support of $Z\mathbf{x}_\phi$ is equal to the support of

$$[E_\phi \ \cdots \ E_\phi]$$

and the absolute value of $Z\mathbf{x}_\phi$ on its support equals 1.

The support of $Z(S_N^n \mathbf{x}_\phi)$ is equal to the support of

$$[E_\phi S^{-k} \ \cdots \ E_\phi S^{-k}],$$

where $k \equiv n \mod L$, proving the following result.

Theorem 14.1 *For* $0 \leq n, m < M$,

$$Z\left(S_N^n \mathbf{x}_\phi\right) \quad and \quad Z\left(S_N^m \mathbf{x}_\phi\right),$$

have disjoint supports if and only if $n \not\equiv m \bmod L$.

A *closely spaced shift echo* is any echo of the form

$$\mathbf{v} = \sum_{n=n_0}^{n_0+L-1} a_n S_N^n \mathbf{x}_\phi.$$

The collection of shifts of a closely spaced shift echo is called a collection of *closely spaced shifts*.

Suppose we know that we are looking at a closely spaced shift echo and n_0 is known. Because the collection

$$\{S_N^n \mathbf{x}_\phi : n_0 \leq n < n_0 + L\}$$

is orthogonal, the coefficients of $\mathbf{v}$ can be computed by inner products. However, by Theorem 14.1, we also know that the ZS representation of $\mathbf{v}$,

$$Z\mathbf{v} = \sum_{n=n_0}^{n_0+L-1} a_n Z\left(S_N^n \mathbf{x}_\phi\right),$$

is a linear combination over the disjoint sets

$$Z\left(S_N^n \mathbf{x}_\phi\right), \qquad n_0 \leq n < n_0 + L,$$

and the absolute value of $Z\mathbf{v}$ on $Z\left(S_N^n \mathbf{x}_\phi\right)$ is constant and equals $|a_n|$. The problem of computing the coefficients of the echo can be approached by image processing methods. In the presence of severe non-Gaussian noise, a combined matched filter and image processing approach can lead to a more accurate computation of the echo's coefficients.

In numerical experiments we consider cyclic echoes of modulated permutation sequences. In this case the echoes are a collection of shifted algebraic lines.

Example 14.3 For $T = 50$, $\gamma = 1/2$, set $\nu = 10$. Then $\gamma T^2 = 1250$. Choosing $N = 625$, the subsampled discrete chirp $\mathbf{x}$ is the modulated permutation sequence μ_{24}. Figure 14.1 displays two of the shifted chirps with scattering coefficients and Figure 14.2 displays their sum. Figures 14.3 and 14.4 are the corresponding ZS representations.

14.2 Fourier Transform of Zero-Padded Vectors

For the rest of this chapter, L and $J > 1$ are integers, $v = e^{2\pi i \frac{1}{L}}$, $\rho = e^{2\pi i \frac{1}{JL}}$,

$$(0.0557 - 1.3493i)S_{25}^5\mathbf{x}$$

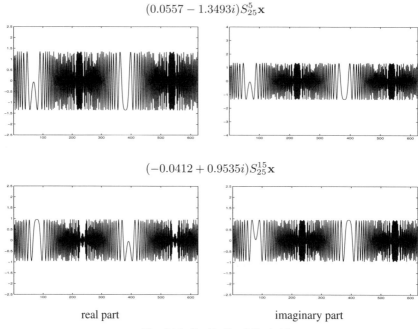

$$(-0.0412 + 0.9535i)S_{25}^{15}\mathbf{x}$$

real part   imaginary part

Fig. 14.1. Cyclically shifted chirp

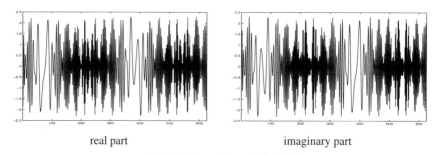

real part   imaginary part

Fig. 14.2. Sum of two shifted chirps

$$F = F(JL), \quad D = D_{JL}, \quad S = S_{JL},$$

$$P(JL, L) \quad JL \times JL,$$

$$D_L(JL) = D\left([\rho^n]_{0 \le n < L}\right),$$

$$C = C_L(JL) = F(L)D_L(JL)F(L)^{-1},$$

$$(0.0557 - 1.3493i)S_{25}^5\mathbf{x}$$

$$(-0.0412 + 0.9535i)S_{25}^{15}\mathbf{x}$$

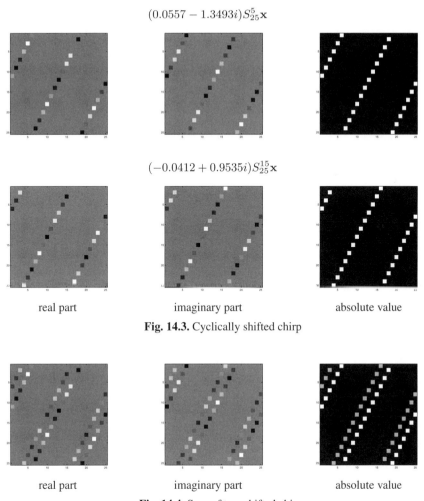

real part   imaginary part   absolute value

Fig. 14.3. Cyclically shifted chirp

real part   imaginary part   absolute value

Fig. 14.4. Sum of two shifted chirps

and

$$\mathcal{C} = \begin{bmatrix} I_L \\ C \\ \vdots \\ C^{J-1} \end{bmatrix}, \qquad \mathcal{D} = D_J(JL) \otimes D_L.$$

We extend results on the Fourier transform of zero-padded periodic sequences to shifts of zero-padded periodic sequences. The main idea is to embed the linear shift model in a larger size cyclic shift model.

Suppose $x(n)$, $n \in \mathbb{Z}$, is a complex sequence. The *linear shift* operator T on complex sequences is defined by

$$(Tx)(n) = x(n-1), \qquad n \in \mathbb{Z}.$$

Aperiodic correlation is defined using linear shifts.

Suppose $\mathbf{u} \in \mathbb{C}^L$. By Theorem 4.6,

$$F\mathbf{u}^J = P\mathcal{C}F(L)\mathbf{u},$$

and

$$P^{-1}F\mathbf{u}^J = \begin{bmatrix} F(L)\mathbf{u} \\ CF(L)\mathbf{u} \\ \vdots \\ C^{J-1}F(L)\mathbf{u} \end{bmatrix} = \mathcal{C}F(L)\mathbf{u}$$

from which we see that the vector formed from the first L components of $P^{-1}F\mathbf{u}^J$ is $F(L)\mathbf{u}$ and the vectors formed from the remaining contiguous blocks of L components are the filtered versions of $F(L)\mathbf{u}$,

$$CF(L)\mathbf{u}, \quad \ldots, \quad C^{J-1}F(L)\mathbf{u}.$$

We extend this result to the shifts

$$S^m \mathbf{u}^J, \qquad 0 \le m < JL.$$

$$FS^m \mathbf{u}^J = D^m F\mathbf{u}^J = D^m P\mathcal{C}F(L)\mathbf{u} = P\left(P^{-1}D^m P\right)\mathcal{C}F(L)\mathbf{u}.$$

To compute the diagonal matrix $P^{-1}DP$ we apply P^{-1} to the vector formed by the diagonal entries of D. Write

$$D = D\left([\rho^m]_{0 \le m < JL}\right), \qquad \rho = e^{2\pi i \frac{1}{JL}}.$$

$P^{-1} = P(JL, J)$. Striding through the vector $[\rho^m]_{0 \le m < JL}$ by J and using $\rho^J = e^{2\pi i \frac{1}{L}}$, we have

$$P^{-1}DP = \begin{bmatrix} D_L & & & & \\ & \rho D_L & & & \\ & & \ddots & & \\ & & & & \\ & & & & \rho^{J-1}D_L \end{bmatrix} = D_J(JL) \otimes D_L.$$

Placing this result into the formula for $FS^n \mathbf{u}^J$, we have the following result.

Suppose $0 \le m < JL$ and

$$m = n + sL, \qquad 0 \le n < L, \, 0 \le s < J.$$

Theorem 14.2

$$P^{-1}F\left(S^m \mathbf{u}^J\right) = \left(D_J^m(JL) \otimes D_L^n\right)\mathcal{C}F(L)\mathbf{u} = \mathcal{D}_M\mathcal{C}F(L)\mathbf{u}.$$

The vector in $\mathbb{C}^L$ formed from the first L components of

$$P^{-1}F\left(S^m\mathbf{u}^J\right)$$

is

$$F(L)S_L^n\mathbf{u} = D_L^nF(L)\mathbf{u}.$$

Example 14.4 $F = F(3L)$, $P^{-1} = P(3L, 3)$ and

$$C = F(L)D_L(3L)F(L)^{-1}.$$

Then for $\mathbf{u} \in \mathbb{C}^L$

$$P^{-1}F\left(\mathbf{u}^3\right) = \begin{bmatrix} F(L)\mathbf{u} \\ CF(L)\mathbf{u} \\ C^2F(L)\mathbf{u} \end{bmatrix},$$

$$P^{-1}F\left(S_{3L}\mathbf{u}^3\right) = \begin{bmatrix} D_LF(L)\mathbf{u} \\ \rho D_LCF(L)\mathbf{u} \\ \rho^2 D_LC^2F(L)\mathbf{u} \end{bmatrix},$$

and for $m = n + sL$, $0 \le n < L$, $0 \le s < 3$,

$$P^{-1}F\left(S_{3L}^m\mathbf{u}^3\right) = \begin{bmatrix} D_L^nF(L)\mathbf{u} \\ \rho^m D_L^nCF(L)\mathbf{u} \\ \rho^{2m} D_L^nC^2F(L)\mathbf{u} \end{bmatrix}, \qquad \rho = e^{2\pi i \frac{1}{3L}}.$$

We see from the example that the vector formed from the first L components of

$$P^{-1}F\left(S_{3L}^m\mathbf{u}^3\right)$$

is

$$F(L)S_L^n\mathbf{u}^3 = D_L^nF(L)\mathbf{u}^3, \qquad 0 \le m < 3L.$$

14.3 ZS Representation of Zero-Padded Vectors

Set

$$M = M_{JL\times K}, \qquad Z = Z_{JL\times K}.$$

Suppose $\mathbf{x} \in \mathbb{C}^N$,

$$M_L\mathbf{x} = \begin{bmatrix} \mathbf{x}_0 & \cdots & \mathbf{x}_{K-1} \end{bmatrix} \text{ and } Z_L\mathbf{x} = \begin{bmatrix} X_0 & \cdots & X_{K-1} \end{bmatrix}.$$

$$M\mathbf{x}^J = \begin{bmatrix} \mathbf{x}_0^J & \cdots & \mathbf{x}_{K-1}^J \end{bmatrix}.$$

Applying the formula for the finite Fourier transform of zero-padded vectors, we have the next result.

Theorem 14.3

$$P^{-1}Zx^J = \begin{bmatrix} Z_L x \\ CZ_L x \\ \vdots \\ C^{J-1} Z_L x \end{bmatrix} = \mathcal{C} Z_L x.$$

Example 14.5 For $L = 15$ and $J = 2$, Figure 14.5 displays Ze_{19}^2 and $P^{-1}Ze_{19}^2$ and compares $P^{-1}Ze_{19}^2$, $P^{-1}Ze_{21}^2$ and $P^{-1}Ze_{36}^2$.

$$Ze_{19}^2 \qquad\qquad\qquad\qquad P^{-1}Ze_{19}^2$$

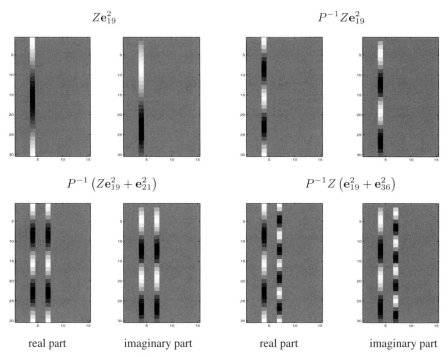

$$P^{-1}\left(Ze_{19}^2 + e_{21}^2\right) \qquad\qquad P^{-1}Z\left(e_{19}^2 + e_{36}^2\right)$$

real part   imaginary part   real part   imaginary part

Fig. 14.5. Zak transform of zero-padded vector

A subset

$$\Sigma \subset \mathbb{Z}/JL \times \mathbb{Z}/K$$

is called a *window* of $JL \times K$ Zak space. If

$$F : \mathbb{Z}/JL \times \mathbb{Z}/K \longrightarrow \mathbb{C},$$

then the restriction of F to Σ is called the *image* of F over the window Σ. By abuse of language, if $\mathbf{y} \in \mathbb{C}^{NJ}$ then the image of $P^{-1}Z\mathbf{y}$ over the window Σ is called the image of $\mathbf{y}$ over the window Σ.

Define the windows Σ_j, $0 \leq j < J$, of $JL \times K$ ZS by

$$\Sigma_j = \{(n, k) : jL \leq n < (j+1)L, \ 0 \leq k < K\}.$$

By Theorem 14.3 the image of $\mathbf{x}^J$ over Σ_0 is

$$Z_L \mathbf{x},$$

and the image of $\mathbf{x}^J$ over Σ_j is the filtered version of $Z_L \mathbf{x}$,

$$C^j Z_L \mathbf{x}, \quad 1 \leq j < J.$$

Because

$$M\left(S_{JN}\mathbf{x}^J\right) = \begin{bmatrix} S\mathbf{x}_{K-1}^J & \mathbf{x}_0^J & \cdots & \mathbf{x}_{K-2}^J \end{bmatrix},$$

we have by Theorem 14.2

$$P^{-1}Z\left(S_{JN}\mathbf{x}^J\right) = \begin{bmatrix} \mathcal{D}CX_{K-1} & CX_0 & \cdots & CX_{K-2} \end{bmatrix}.$$

The image of $S_{JN}\mathbf{x}^J$ over Σ_0 is

$$Z_L\left(S_N\mathbf{x}\right) = \begin{bmatrix} D_L X_{k-1} & X_0 & \cdots & X_{K-2} \end{bmatrix}$$

and the image of $S_{JN}\mathbf{x}^J$ over Σ_j is

$$\begin{bmatrix} \rho^j D_L C^j X_{K-1} & C^j X_0 & \cdots & C^j X_{K-2} \end{bmatrix}, \quad 1 \leq j < J, \ \rho = e^{2\pi i \frac{1}{JL}}.$$

Continuing in this way we have the following result.

Theorem 14.4 *For $0 \leq k < K$*

$$P^{-1}Z\left(S_{JN}^k \mathbf{x}^J\right) = \begin{bmatrix} \mathcal{D}CX_{K-k} & \cdots & \mathcal{D}CX_{K-1} & CX_0 & \cdots & CX_{K-k-1} \end{bmatrix}$$

and for $0 \leq n < JL$

$$P^{-1}Z\left(S_{JN}^{k+nK} \mathbf{x}^J\right) = \mathcal{D}^n P^{-1}Z\left(S_{JN}^k \mathbf{x}^J\right).$$

In Table 14.1, $m = t + sK$, $0 \leq t < K$, $0 \leq s < JL$.

Since the image of $S_{JN}^m \mathbf{x}^J$ over the window Σ_0 is

$$Z_L\left(S_N^m \mathbf{x}\right),$$

for $m_1 \equiv m_2 \mod N$, we cannot distinguish the images of

$$S_{JN}^{m_1} \mathbf{x}^J \text{ and } S_{JN}^{m_2} \mathbf{x}^J$$

over the window Σ_0. Tables 14.2 and 14.3 detail the image over the window Σ_j and image of $S_{JN}^{sK} \mathbf{x}^J$.

Table 14.1. The image over the window Σ_0

	Σ_0
$\mathbf{x}^J$	$X_0 \quad \ldots \quad X_{K-1}$
$S_{JN}^t \mathbf{x}^J$	$D_L X_{K-t} \quad \ldots \quad D_L X_{K-1} \quad X_0 \quad \ldots \quad X_{K-t-1}$
$S_{JN}^{sK} \mathbf{x}^J$	$D_L^s X_0 \quad \ldots \quad D_L^s X_{K-1}$
$S_{JN}^m \mathbf{x}^J$	$D_L^s \left[D_L X_{K-t} \quad \ldots \quad D_L X_{K-1} \quad X_0 \quad \ldots \quad X_{K-t-1} \right]$

Table 14.2. The image over the window Σ_j

	Σ_j
$\mathbf{x}^J$	$C^j X_0 \quad \ldots \quad C^j X_{K-1}$
$S_{JN}^t \mathbf{x}^J$	$\rho^j D_L C^j X_{K-t} \quad \ldots \quad \rho^j D_L C^j X_{K-1} \quad C^j X_0 \quad \ldots \quad C^j X_{K-t-1}$
$S_{JN}^{sK} \mathbf{x}^J$	$\rho^{js} D_L^s \left[C^j X_0 \quad \ldots \quad C^j X_{K-1} \right]$
$S_{JN}^m \mathbf{x}^J$	$\rho^{js} D_L^s \left[\rho^j D_L C^j X_{K-t} \quad \ldots \right.$ $\left. \rho^j D_L C^j X_{K-1} \quad C^j X_0 \quad \ldots \quad C^j X_{K-t-1} \right]$

Table 14.3. The image of $S_{JN}^{sK} \mathbf{x}^J$

	$S_{JN}^{sK} \mathbf{x}^J$
Σ_0	$D_L^s X_0 \quad \ldots \quad D_L^s X_{K-1}$
Σ_1	$\rho^s D_L^s C X_0 \quad \ldots \quad \rho^s D_L^s C X_{K-1}$
$\vdots$	
Σ_j	$\rho^{js} D_L^s C^j X_0 \quad \ldots \quad \rho^{js} D_L^s C^j X_{K-1}$
$\vdots$	
Σ_{J-1}	$\rho^{(J-1)s} D_L^s C^{J-1} X_0 \quad \ldots \quad \rho^{(J-1)s} D_L^s C^{J-1} X_{K-1}$

14.4 Modulated Permutation Sequences

$N = LK = JL^2$, where $L > 1$, $J > 1$ are integers. $\Lambda = Perm(L)$.

Suppose $\mathbf{x}_\phi$, $\phi \in \Lambda$, is a modulated permutation sequence in $\mathbb{C}^N$.

$$Z_L \mathbf{x}_\phi = [E_\phi \quad \ldots \quad E_\phi] D(\mathbf{x}),$$

where $\mathbf{x} \in \mathbb{C}_1^K$ and E_ϕ is repeated J times. The discussion includes any unit discrete chirp in $\mathbb{C}^N$, but for the present we do not assume that $\mathbf{x}_\phi$ satisfies ideal autocorrelation.

Consider the zero-padded vector $\mathbf{x}_\phi^J$ of $\mathbf{x}_\phi$ in $\mathbb{C}^{JN}$. In general

$$\mathbf{x}_\phi^J \quad \text{and} \quad S_{JN} \mathbf{x}_\phi^J$$

are not orthogonal. However, as

$$S_{JN}^{j}\mathbf{x}_{\phi}^{J} \text{ and } S_{JN}^{j+N}\mathbf{x}_{\phi}^{J}, \qquad 0 \le j < JN,$$

have disjoint supports, they must be orthogonal.

By Theorems 14.3 and 14.4, the images of $\mathbf{x}_{\phi}^{J}$ and $S_{JN}\mathbf{x}_{\phi}^{J}$ over the window Σ_0 are

$$Z_L(\mathbf{x}_{\phi}) \text{ and } Z_L(S_N\mathbf{x}_{\phi}).$$

$Z_L(\mathbf{x}_{\phi})$ and $Z_L(S_N\mathbf{x}_{\phi})$ have disjoint supports, even though $\mathbf{x}_{\phi}^{J}$ and $S_{JN}\mathbf{x}_{\phi}^{J}$ are not orthogonal.

Theorem 14.5 *The images over the window Σ_0 of the sequences*

$$S_{JN}^{n}\mathbf{x}_{\phi}^{J}, \qquad 0 \le n < L,$$

have pairwise disjoint supports.

Consider an echo

$$\mathbf{v} = \sum_{n=0}^{L-1} a_n S_{JN}^{n}\mathbf{x}_{\phi}^{J}$$

and its ZS representation

$$Z\mathbf{v} = \sum_{n=0}^{L-1} a_n Z\left(S_{JN}^{n}\mathbf{x}_{\phi}^{J}\right).$$

Windowing over Σ_0 the expansion becomes

$$\sum_{n=0}^{L-1} a_n Z_L\left(S_N^{n}\mathbf{x}_{\phi}^{J}\right).$$

As discussed in Section 14.1, the coefficients can be computed by image processing or inner products over $L \times K$ ZS. The argument easily generalizes to echoes of the form

$$\mathbf{v} = \sum_{n=n_0}^{n_0+L-1} a_n S_{JN}^{n}\mathbf{x}_{\phi}^{J}.$$

However, n_0 must be known, as windowing over Σ_0 is invariant under S_{JN}^{N}.

Example 14.6 For $L = 15$, $J = 2$, Figure 14.6 displays $P^{-1}Z\mathbf{x}_{\mu_2}^{2}$ and $P^{-1}Z\left(S^4\mathbf{x}_{\mu_2}^{2}\right)$. Figure 14.8 compares $P^{-1}Z\left(S^{19}\mathbf{x}_{\mu_2}^{2}\right)$ and $P^{-1}Z\left(S^{34}\mathbf{x}_{\mu_2}^{2}\right)$. Figures 14.7 and 14.9 show the nonzero values over the window Σ_0.

$$P^{-1}Z\mathbf{x}_{\mu_2}{}^2 \qquad\qquad P^{-1}Z\left(S^4\mathbf{x}_{\mu_2}{}^2\right)$$

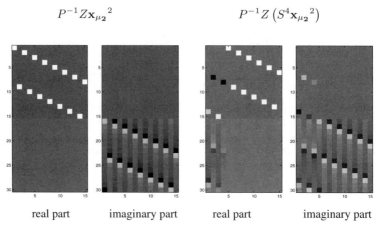

real part   imaginary part   real part   imaginary part

Fig. 14.6. Zak transform of zero-padded permutation sequence

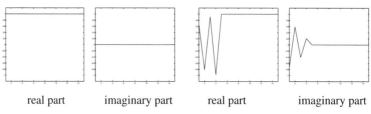

real part   imaginary part   real part   imaginary part

Fig. 14.7. Nonzero values over the window Σ_0

Suppose $\mathbf{x}_\phi$ satisfies ideal autocorrelation. Over the window Σ_0 the images of

$$S_{JN}^n\mathbf{x}_\phi^J, \qquad 0 \le n < N,$$

are orthogonal. Consider an echo of the form

$$\mathbf{v} = \sum_{n=0}^{N-1} a_n S_{JN}^n\mathbf{x}_\phi^J.$$

Over the window Σ_0 we have the orthogonal expansion

$$\sum_{n=0}^{N-1} a_n Z_L\left(S_N^n\mathbf{x}_\phi\right).$$

The coefficients can be computed by inner products over Σ_0 or by a combined image processing and inner product approach. The argument easily generalizes to echoes of the form

$$\sum_{n=n_0}^{n_0+N-1} a_n S_{JN}^n\mathbf{x}_\phi^J,$$

$$P^{-1}Z\left(S^{4+15}\mathbf{x}_{\mu_2}{}^2\right) \qquad\qquad P^{-1}Z\left(S^{4+30}\mathbf{x}_{\mu_2}{}^2\right)$$

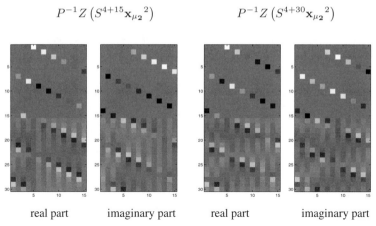

real part imaginary part real part imaginary part

Fig. 14.8. Zak transform of zero-padded permutation sequence

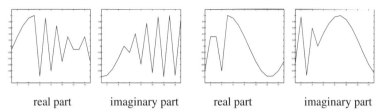

real part imaginary part real part imaginary part

Fig. 14.9. Nonzero values over the window Σ_0

with the same caution about n_0.

The echoes studied in this section are called *closely spaced shifted echoes*. If the echo **v** can be decomposed into a collection of closely spaced shift echoes, then these methods can be used to determine the coefficients of **v**. This is the case, even though the shifts associated to a closely spaced shift echo are never orthogonally related to each other.

14.5 Global Computations

$v = e^{2\pi i\frac{1}{L}}$ and $\rho = e^{2\pi i\frac{1}{JL}}$. The notation is as in the preceding section.

Windowing does not give complete information about a sequence and its shifts, especially orthogonality information. For example, a collection of closely spaced shifts of a zero-padded discrete chirp is not orthogonal, but its projection as viewed through the ZS window Σ_0 is orthogonal. Complete information requires windowing over the whole of ZS. The major obstacle to directly using complete ZS information is that D_L^m and C^j do not commute. In this section we derive a formula for describing this noncommutativity. This formula will be used in the next section to describe the orthogonal properties of zero-padded discrete chirps and their shifts.

Some preliminary formulas are required. Set $\mathbf{e}_s = \mathbf{e}_s^L$ and consider the $L \times L$ diagonal matrices $D(\mathbf{e}_s)$, $0 \le s < L$. Write

$$D_L(JL) = \sum_{s=0}^{L-1} \rho^s D_L(\mathbf{e}_s).$$

Since

$$F(L)D(\mathbf{e}_0)F(L)^{-1} = I(L, L),$$

we have

$$F(L)D(\mathbf{e}_s)F(L)^{-1} = F(L)S_L^s D(\mathbf{e}_0) S_L^{-s} F(L)^{-1}$$
$$= D_L^s I(L, L)D_L^{-s},$$

where $I(L, L)$ is the $L \times L$ matrix, all of whose coefficients are 1. Set

$$I_s(L, L) = D_L^s I(L, L)D_L^s.$$

$I_s(L, L)$ is an $L \times L$ circulant matrix whose 0-th column is F_s, the s-th column of $F(L)$. Then

$$C^j = F(L)D_L^j(JL)F(L)^{-1} = \frac{1}{L}\sum_{s=0}^{L-1} \rho^{js} I_s(L, L).$$

Set

$$m = n + sL, \quad 0 \le n < L, \ 0 \le s < J.$$

Theorem 14.6

$$\rho^{mj} D_L^m C^j = \rho^{jsL}\left(C^j + \frac{\alpha_j}{L}\sum_{s=0}^{n-1} \rho^{js} I_s(L, L)\right) D_L^n,$$

where $\alpha_j = \rho^{jL} - 1$, $0 \le j < J$.

Proof

$$\rho^{km} D_L^m C^j = \frac{1}{L}\rho^{jsL}\rho^{jn}\left(\sum_{s=0}^{L-1} \rho^{js} I_{s+n}(L, L)\right) D_L^n.$$

By the change of variables $t = s + n$,

$$\rho^{jn}\sum_{s=0}^{L-1} \rho^{js} I_{s+n}(L, L) = \sum_{t=n}^{L-1+n} \rho^{jt} I_t(L, L)$$

$$= \sum_{t=L}^{L-1+n} \rho^{jt} I_t(L, L) + \sum_{t=n}^{L-1} \rho^{jt} I_t(L, L).$$

By the change of variables $s = t - L$ in the first summation,

$$\sum_{t=L}^{L-1+n} \rho^{jt} I_t(L,L) = \rho^{jL} \sum_{s=0}^{n-1} \rho^{js} I_s(L,L)$$

$$= \left(\rho^{jL} - 1\right) \sum_{s=0}^{n-1} \rho^{js} I_s(L,L) + \sum_{s=0}^{n-1} \rho^{js} I_s(L,L).$$

Then

$$\rho^{jn} \sum_{s=0}^{L-1} \rho^{js} I_{s+n}(L,L) = \sum_{s=0}^{L-1} \rho^{js} I_s(L,L) + \left(\rho^{jL} - 1\right) \sum_{s=0}^{n-1} \rho^{js} I_s(L,L)$$

$$= LC^j + \left(\rho^{jL} - 1\right) \sum_{s=0}^{n-1} \rho^{js} I_s(L,L),$$

completing the proof.

We also want a formula for

$$\rho^{mj} C^{-j} D_L^m C^j$$

for computing inner products. Because

$$D_L^s I(L,L) = [F_s \ \cdots \ F_s] = F(L) [E_s \ \cdots \ E_s] ,$$

we have

$$C^{-j} I_s(L,L) = F(L) D_L^{-j} (JL) F(L)^{-1} [F_s \ \cdots \ F_s] D_L^{-s}$$
$$= F(L) D_L^{-j} (JL) [E_s \ \cdots \ E_s] D_L^{-s}$$
$$= \rho^{-js} F(L) [E_s \ \cdots \ E_s] D_L^{-s}$$
$$= \rho^{-js} D_L^s I(L,L) D_L^{-s},$$

proving the next theorem.

Theorem 14.7

$$\rho^{mj} C^{-j} D_L^m C^j = \rho^{sjL} \left(I_L + \frac{\alpha_j}{L} \sum_{s=0}^{n-1} I_s(L,L) \right) D_L^n, \quad 0 \le j < J,$$

where $\alpha_j = \rho^{jL} - 1$.

14.6 Orthogonality

$N = LK = L^2 J$, $L > 1$, $J \ge 1$ are integers. $\Lambda = Perm(L)$.

The formula of Theorem 14.3 directly applies to computing inner products of a zero-padded sequence and its shifts. Suppose e_ϕ, $\phi \in \Lambda$, is a permutation sequence in $\mathbb{C}^N$ and e_ϕ^J its zero-padding in $\mathbb{C}^{JN}$.

Theorem 14.8 *The collection of shifts*

$$\{S_{JN}^{mK} \mathbf{e}_\phi^J : 0 \leq m < JL\}$$

is orthogonal.

Proof We show that

$$Z\mathbf{e}_\phi^J \text{ and } Z\left(S_{JN}^{mK}\mathbf{e}_\phi^J\right), \quad 1 \leq m < JL,$$

are orthogonal. A typical column vector in $P^{-1}Z\mathbf{e}_\phi^J$ is

$$C^j E_{\phi(t)}, \quad 0 \leq j < J, 0 \leq t < K.$$

The corresponding column vector in $P^{-1}Z\left(S_{JN}^{mK}\mathbf{e}_\phi^J\right)$ is

$$\rho^{mj} D_L^m C^j E_{\phi(t)}.$$

The inner product of these two column vectors is

$$\langle E_{\phi(t)}, \rho^{mj} C^{-j} D_L^m C^j E_{\phi(t)}\rangle,$$

which by Theorem 14.6 is

$$\rho^{-sjL} v^{-n\phi(t)} \left(1 + n\frac{\alpha_j^*}{L}\right).$$

Summing over $0 \leq j < J$ and $0 \leq t < L$,

$$\sum_{t=0}^{K-1} v^{-n\phi(t)} \sum_{j=0}^{J-1} \rho^{-sjL}\left(1 + n\frac{\alpha_j^*}{L}\right).$$

If $n \neq 0$, the first sum is 0. Otherwise the sum is

$$K \sum_{j=0}^{J-1} \rho^{-sjL} = 0,$$

because $m \neq 0$, completing the proof.

We have mentioned that

$$\mathbf{e}_\phi^J \text{ and } S_{JN}\mathbf{e}_\phi^J$$

are not orthogonal. Specifically, we have the following result.

Theorem 14.9

$$\langle \mathbf{e}_\phi^J, S_{JN}\mathbf{e}_\phi^J\rangle = -\frac{1}{L^2} v^{-\phi(L-1)}.$$

Proof   The inner product is the sum over $0 \le j < J$ of the vectors

$$E_{\phi(0)} \text{ and } \rho^j C^{-j} D_L C^j E_{\phi(L-1)}.$$

By Theorem 14.6, with $n = 1$ and $s = 0$,

$$\rho^j C^{-j} D_L C^j = D_L + \frac{\alpha_j}{L} I(L, L) D_L.$$

Then

$$\rho^j C^{-j} D_L C^j E_{\phi(L-1)} = v^{\phi(L-1)} E_{\phi(L-1)} + \frac{\alpha_j}{L} v^{\phi(L-1)} \mathbf{1},$$

and the ZS inner product is

$$\frac{\alpha_j^*}{L} v^{-\phi(L-1)}.$$

Summing over $0 \le j < J$, the ZS inner product of

$$\mathbf{e}_\phi^J \text{ and } S_{JN} \mathbf{e}_\phi^J$$

is

$$-\frac{J}{L} v^{-\phi(L-1)},$$

completing the proof.

15

Sequence Shaping

The mathematical properties of the sequences, sequence pairs and sequence sets designed in this text, except for a few, are governed by periodic autocorrelations and cross correlations. Two sequence sets having the same periodicity and size and satisfying pairwise ideal correlation cannot be distinguished mathematically. However, mathematically indistinct sequence sets can have different properties when used in applications. For example, as described in the text, their acyclic correlation properties can vary. In this chapter we discuss modifications of the general theory to achieve two application requirements, limiting the bandwidth of sequences and constructing real-valued sequences.

15.1 Bandlimiting

In this section we discuss two standard methods for manipulating bandwidth, interpolation and bandpass filtering, and we provide examples of the effects of these signal shaping methods on acyclic correlation properties. The key observation is the dependence of the quality of the resulting sequences on the input sequences. In general, interpolation increases the width of the autocorrelation mainlobe, while bandpass filtering degrades the polyphase property.

For $L = 87$, $R = 1$, the permutation sequence pair $(\mathbf{x}_{\phi_0}, \mathbf{x}_{\mu_2})$, satisfies ideal correlation. Figures 15.1 and 15.2 display the results of bandpass filtering the sequences. In detail, portions of the Fourier transform of the sequences are set to zero, then the inverse Fourier transform is applied to generate the bandlimited sequences. The acyclic autocorrelations and the acyclic cross correlation are plotted on the same axis as amplitude decibel plots. The absolute values of the Fourier transform of either of the resulting sequences are identical. 50% and 80% of the Fourier transform coefficients have been set to zero, respectively. The resulting sequences have very much the same shape, and only one of the pair is displayed.

For $L = 61$, $R = 1$, the permutation sequence pair $(\mathbf{x}_{\phi_0}, \mathbf{x}_{\mu_2})$, satisfies ideal correlation. Figure 15.3 displays the results of interpolating the sequences. Cubic

M. An et al., *Ideal Sequence Design in Time-Frequency Space*,
DOI 10.1007/978-0-8176-4738-4_15,

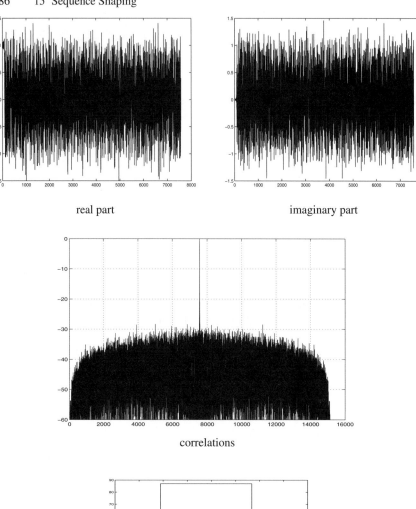

real part   imaginary part

correlations

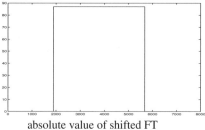

absolute value of shifted FT

Fig. 15.1. Bandlimited sequences, $L = 87$, $N = 7569$

polynomial interpolation is used to fit 2 values to each value in the sequence. The resulting sequences are of length 7441.

Figure 15.4 displays the results of a similar experiment with $L = 39$ and interpolating to fit 5 values to each value in the sequence. The resulting sequence is of length 7601.

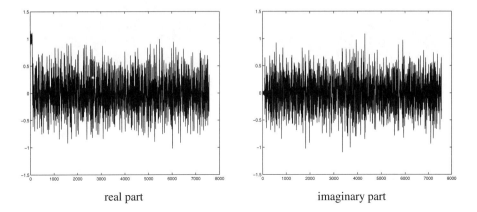

real part   imaginary part

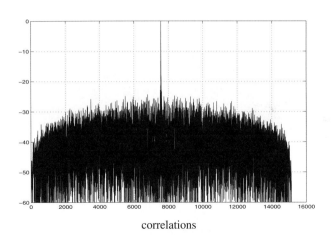

correlations

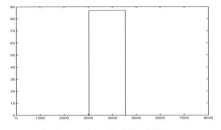

absolute value of shifted FT

Fig. 15.2. Bandlimited sequences, $L = 87$, $N = 7569$

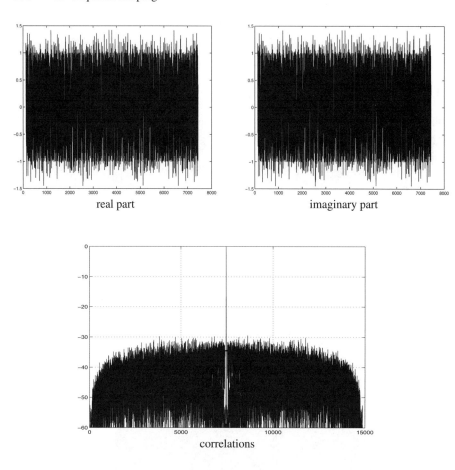

real part   imaginary part

correlations

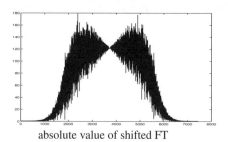

absolute value of shifted FT

Fig. 15.3. Interpolated sequences, $L = 61$, $N = 7441$

For $R = 3$, we have the pair of discrete chirps satisfying ideal correlation, $\mathbf{x}_1$ and $\mathbf{x}_2$,

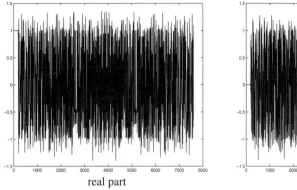
real part

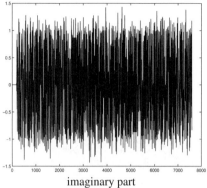
imaginary part

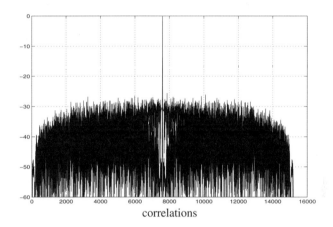
correlations

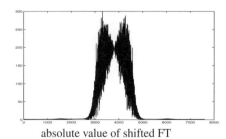
absolute value of shifted FT

Fig. 15.4. Interpolated sequences, $L = 39$, $N = 7601$

$$\mathbf{x}_1 = \begin{bmatrix} 1 \\ w \\ 1 \end{bmatrix}, \quad \mathbf{x}_1 = \begin{bmatrix} 1 \\ w \\ w \end{bmatrix}, \quad w = e^{2\pi i \frac{1}{3}}.$$

The discrete carrier frequency for $\mathbf{x}_1$ is $\frac{1}{2}$ and that for $\mathbf{x}_2$ is 0. Set

$$\mathbf{a}_r(s) = \mathbf{x}_1, \qquad \mathbf{b}_r(s) = \mathbf{x}_2, \qquad 0 \le s < L.$$

For $L = 25$, the modulated permutation sequences are of length 1875. Figures 15.5 and 15.6 display the interpolated sequences and interpolated, bandlimited sequences of length 7497.

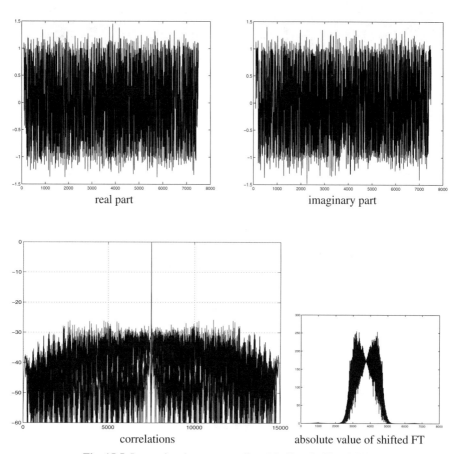

real part   imaginary part

correlations   absolute value of shifted FT

Fig. 15.5. Interpolated sequences, $L = 25$, $R = 3$, $N = 7497$

15.2 Real-Valued Sequences

Many sonar and radar applications require real-valued sequences. In the following discussion, we present two methods for modifying the polyphase sequences designed in the preceding chapters into real-valued sequences. The second method forms the sequence by simply taking the real part of the polyphase sequence. Both methods

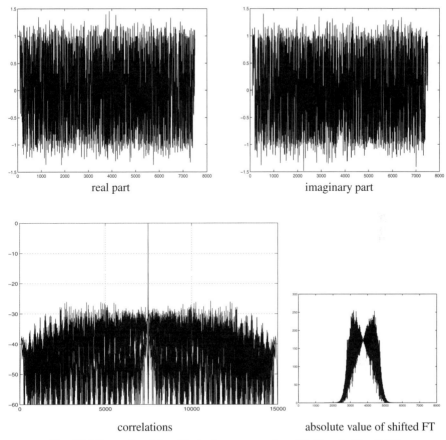

real part   imaginary part

correlations   absolute value of shifted FT

Fig. 15.6. Interpolated, bandlimited sequences, $L = 25$, $R = 3$, $N = 7497$

will be described in Zak space and their correlation properties will be compared using the Zak space (ZS) correlation formula. This has the advantage that the echo analysis methodologies of Chapter 14 continue to hold with slight modifications. The discussion is limited to permutation sequences. Set $\Lambda = Perm(L)$ and $N = L^2$.

Suppose $\phi \in \Lambda$ and $\mathbf{e}_\phi$ is the corresponding permutation sequence. Consider the *time-reversal permutation* $\rho \in \Lambda$ defined by

$$\rho(r) = L - r, \quad 0 \le r < L.$$

For L odd ρ is a $*$-permutation with

$$\phi_0 - \rho = \mu_2.$$

Set E'_ϕ equal to the $2L \times L$ matrix

$$E'_\phi = \begin{bmatrix} E_\phi \\ E_{\phi\rho} \end{bmatrix}$$

and define $e'_\phi \in \mathbb{C}^{2N}$ by

$$Z_{2L} e'_\phi = E'_\phi.$$

Example 15.1 $L = 5$ and $\phi = (0 \ 2 \ 4 \ 1 \ 3)$. Then $\phi\rho = (0 \ 3 \ 1 \ 4 \ 2)$ and

$$E'_\phi = \begin{bmatrix} 1\ 0\ 0\ 0\ 0 \\ 0\ 0\ 0\ 1\ 0 \\ 0\ 1\ 0\ 0\ 0 \\ 0\ 0\ 0\ 0\ 1 \\ 0\ 0\ 1\ 0\ 0 \\[6pt] 1\ 0\ 0\ 0\ 0 \\ 0\ 0\ 1\ 0\ 0 \\ 0\ 0\ 0\ 0\ 1 \\ 0\ 1\ 0\ 0\ 0 \\ 0\ 0\ 0\ 1\ 0 \end{bmatrix}.$$

Write

$$F(10) = [F_0 \ F_1 \ \cdots \ F_9].$$

$F_0 = \mathbf{1}^{10}$, $F_5 = [(-1)^n]_{0 \le n < 10}$ and

$$F_n = F^*_{-n} = [w^{nm}]_{0 \le m < 10}, \qquad 0 \le n < 10,$$

where $-n$ is taken modulo 10.

$$F(10)^* E'_\phi = [F_0 + F_5 \ \ F_2 + F_2^* \ \ F_4 + F_4^* \ \ F_1 + F_1^* \ \ F_3 + F_3^*]$$

and e'_ϕ is a real-valued sequence. Applying the ZS correlation formula to E'_ϕ, we find that e'_ϕ satisfies ideal autocorrelation.

The general case follows in exactly the same way using the conjugacy relationships between the columns of the Fourier transform matrix, proving the following result.

Theorem 15.1 *For $\phi \in \Lambda$, $e'_\phi \in \mathbb{C}^{2N}$ defined by*

$$Z_{2L} e'_\phi = \begin{bmatrix} E_\phi \\ E_{\phi\rho} \end{bmatrix}$$

is a real-valued vector satisfying ideal autocorrelation.

Suppose $\phi, \delta \in \Lambda$ such that $\gamma = \phi^{-1}\delta$ is a $*$-permutation. Form

$$E'_\phi = \begin{bmatrix} E_\phi \\ E_{\phi\rho} \end{bmatrix}, \qquad E'_\delta = \begin{bmatrix} E_\delta \\ E_{\delta\rho} \end{bmatrix}$$

and set e'_ϕ and e'_δ to be the corresponding real-valued sequences in $\mathbb{C}^{2N}$. Applying the ZS correlation formula as in the proof of Theorem 11.6 to $\mathbf{v} = e'_\phi \circ e'_\delta$, we have, up to ZS modulation by a diagonal matrix where diagonal entries have absolute value 1, that $Z_{2L}\mathbf{v}$ is

$$\begin{bmatrix} E_{\phi\phi_m} \\ E_{\phi\rho\phi_n} \end{bmatrix},$$

where $\phi_m = (\phi_0 - \gamma^{-1})^{-1}$ and

$$\phi_n = (\phi_0 - \rho^{-1}\gamma^{-1}\rho)^{-1}.$$

There are exactly two nonzero entries in each column of $Z_{2L}\mathbf{v}$ and we have

$$|v_n| \leq \frac{1}{L}, \quad 0 \leq n < 2N.$$

Theorem 15.2 *If $\phi, \delta \in \Lambda$ such that $\gamma = \phi^{-1}\delta$ is a *-permutation and $\mathbf{v} = \mathbf{e}'_\phi \circ \mathbf{e}'_\delta$, then*

$$|v_n| \leq \frac{1}{L}, \quad 0 \leq n < 2N.$$

As the optimum cross correlation bound for vectors in $\mathbb{C}^{2N}$ satisfying ideal auto-correlation is $\frac{1}{\sqrt{2L}}$, we give up a factor of $\frac{1}{\sqrt{2}}$ in order to have real-valued sequences.

For any $\mathbf{x} \in \mathbb{C}^N$

$$Z_L\mathbf{x}^* = R_L(Z_L\mathbf{x})^*,$$

where $*$ denotes the complex conjugate and

$$R_L = E_\rho$$

is the time-reversal matrix. In particular for $\phi \in \Lambda$,

$$Z_L(\mathbf{e}_\phi + \mathbf{e}_\phi^*) = E_\phi + E_{\rho\phi}.$$

For simplicity assume L is odd. Set

$$\mathbf{u} = (\mathbf{e}_\phi + \mathbf{e}_\phi^*) \circ (\mathbf{e}_\phi + \mathbf{e}_\phi^*).$$

By the ZS correlation formula we have the next result.

Theorem 15.3 $U_0 = 21^L + 2E_0$ *and for* $1 \leq r < L$

$$U_r = c_r E_{\phi(m)} + c'_r E_{\phi\rho(m)},$$

where $m \equiv 2^{-1}r \bmod L$,

$$c_r = \begin{cases} v^{-\phi(m)}, & 0 \leq m < r, \\ 1, & r \leq m < L, \end{cases}$$

and

$$c'_r = \begin{cases} v^{-\rho\phi(m)}, & 0 \leq m < r, \\ 1, & r \leq m < L, \end{cases} \quad v = e^{2\pi i \frac{1}{L}}.$$

By Theorem 15.3 the real-valued vector in $\mathbb{R}^N$

$$\mathbf{e}_\phi + \mathbf{e}_\phi^*$$

does not satisfy ideal autocorrelation, but because U_r, $1 \le r < L$, has exactly two nonzero components, the components of $\mathbf{u}$ satisfy

$$u_0 = 2 + \frac{2}{L}$$

and

$$|u_n| \le \frac{2}{L}, \qquad 0 < n < N.$$

Suppose $\phi, \delta \in \Lambda$ and $\gamma = \phi^{-1}\delta$ is a $*$-permutation with

$$\Delta_r(\gamma) = \{m_r\}, \qquad 0 \le r < L.$$

$\gamma' = \phi^{-1}\rho\delta$ is also a $*$-permutation. Set

$$\Delta_r(\gamma') = \{n_r\}, \qquad 0 \le r < L.$$

Define $\mathbf{w} \in \mathbb{C}^N$ by

$$\mathbf{w} = \left(\mathbf{e}_\phi + \mathbf{e}_\phi^*\right) \circ \left(\mathbf{e}_\delta + \mathbf{e}_\delta^*\right) = \left(\mathbf{e}_\phi + \mathbf{e}_{\phi\rho}\right) \circ \left(\mathbf{e}_\delta + \mathbf{e}_{\delta\rho}\right).$$

By the ZS correlation formula for $0 \le r < L$

$$W_r = c_r E_{\phi(m_r)} + c_r' E_{\phi\rho(m_r)} + d_r E_{\phi(n_r)} + d_r' E_{\phi\rho(n_r)},$$

where c_r, c_r', d_r and d_r' have absolute value 1. As a result,

$$|w_n| \le \frac{4}{L}, \qquad 0 \le n < N.$$

The preceding two constructions are related, but the corresponding real-valued sequences have different characteristics. The sequences

$$\mathbf{e}_\phi', \qquad \phi \in \Lambda,$$

are periodic modulo $2N$, have ideal autocorrelation with autocorrelation at $n = 0$ equal to 1 and the components of the cross correlation of a pair

$$\left(\mathbf{e}_\phi', \mathbf{e}_\delta'\right), \qquad \phi^{-1}\delta \in \Lambda^*,$$

are bounded by $\frac{1}{L}$.

The sequences

$$\mathbf{e}_\phi + \mathbf{e}_\phi^*, \qquad \phi \in \Lambda,$$

are periodic modulo N, do not satisfy ideal autocorrelation with the autocorrelation at $n = 0$ equal to $2 + \frac{2}{L}$ and are bounded in absolute value otherwise by $\frac{2}{L}$. The components of the cross correlation of a pair

$$\left(\mathbf{e}_\phi + \mathbf{e}_\phi^*, \mathbf{e}_\delta + \mathbf{e}_\delta^*\right), \qquad \phi^{-1}\delta \in \Lambda^*$$

are bounded in absolute value by $\frac{4}{L}$.

The components of the first set of real-valued sequences are bounded in absolute value by $\frac{1}{L}$ and the components of the second set are bounded in absolute value by $\frac{2}{L}$.

The following numerical experiments compare the correlation properties of the two real-valued sequences constructed in the discussion above.

Example 15.2 For $L = 11$, set

$$\phi_1 = (5\ 0\ 3\ 7\ 2\ 8\ 10\ 4\ 1\ 9\ 6),$$

$$\delta_1 = (5\ 3\ 2\ 10\ 1\ 6\ 0\ 7\ 8\ 4\ 9).$$

$\phi_1^{-1}\delta_1 = \mu_2$ is a $*$-permutation. $\mathbf{e}_{\phi_1}'$ and $\mathbf{e}_{\delta_1}'$ are a pair of real-valued sequences of length 242. Figure 15.7 displays the sequences and their autocorrelations and the cross correlation on the same axis.

For $R = 3$, $L = 9$, set

$$\phi_2 = (7\ 1\ 6\ 3\ 2\ 5\ 8\ 4\ 0),$$

$$\delta_2 = (8\ 4\ 7\ 2\ 6\ 1\ 3\ 5\ 0).$$

$\phi_2^{-1}\delta_2 = \mu_2$ is a $*$-permutation. Taking the modulating vectors to be the unit discrete chirps as before, we have a pair of polyphase sequences of length 243 $(\mathbf{e}_{\phi_2}, \mathbf{e}_{\delta_1})$ satisfying ideal correlation. Figure 15.8 displays the autocorrelations and the cross correlation of the real-valued sequences

$$\mathbf{e}_{\phi_2} + \mathbf{e}_{\phi_2}^*, \qquad \mathbf{e}_{\delta_2} + \mathbf{e}_{\delta_2}^*$$

on the same axis.

Example 15.3 Figures 15.9 and 15.10 display the results of bandlimiting the real-valued sequences in Example 15.2. Symmetric points of the Fourier transform are set to zero to preserve the real-valued sequences. Figures 15.11 and 15.12 display the results of the same experiment using the real parts of the polyphase sequences.

Figure 15.13 displays the results of interpolating the sequences. Cubic polynomial interpolation is used to fit 5 values to each value in the sequence. Figure 15.14 displays the results of interpolating the same sequences by a factor of 10.

Figures 15.15 and 15.16 display the results of the same experiment using the real parts of the polyphase sequences.

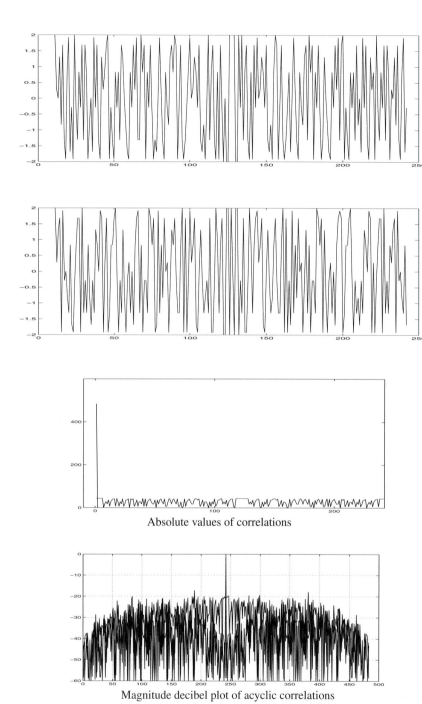

Absolute values of correlations

Magnitude decibel plot of acyclic correlations

Fig. 15.7. Real-valued sequences and their correlations, $N = 242$

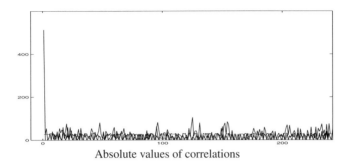

Absolute values of correlations

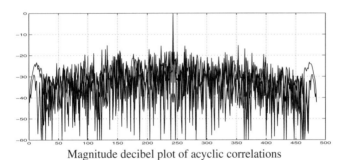

Magnitude decibel plot of acyclic correlations

Fig. 15.8. Correlations of real-valued sequences, $N = 243$

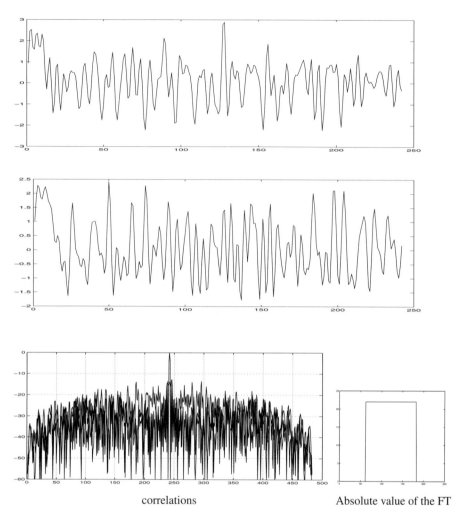

correlations   Absolute value of the FT

Fig. 15.9. Bandlimited real-valued sequences, $N = 242$

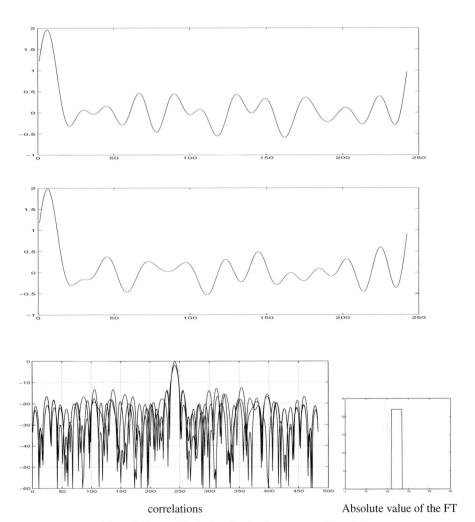

correlations   Absolute value of the FT

Fig. 15.10. Bandlimited real-valued sequences, $N = 242$

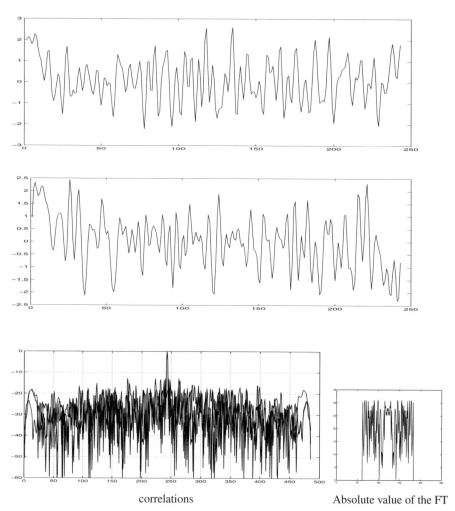

correlations   Absolute value of the FT

Fig. 15.11. Real parts of bandlimited polyphase sequences, $N = 243$

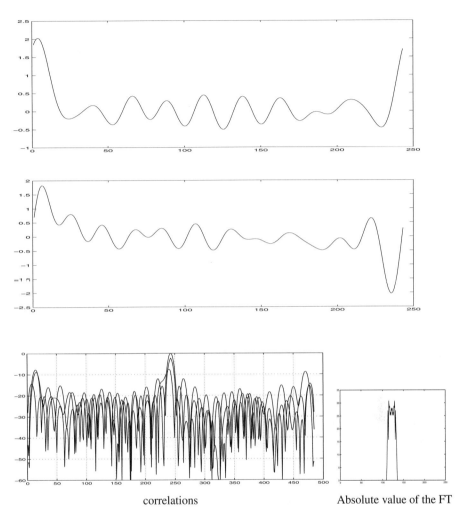

correlations   Absolute value of the FT

Fig. 15.12. Real parts of bandlimited polyphase sequences, $N = 243$

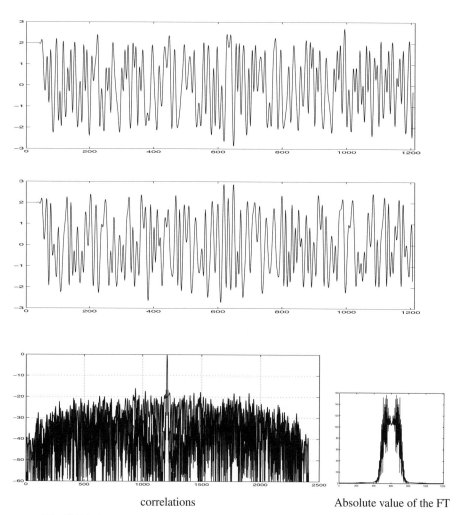

correlations   Absolute value of the FT

Fig. 15.13. Interpolated real-valued sequences and their correlations, $N = 1206$

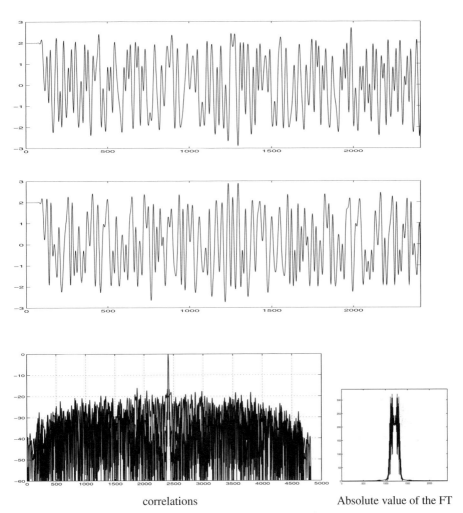

correlations   Absolute value of the FT

Fig. 15.14. Interpolated real-valued sequences and their correlations, $N = 2411$

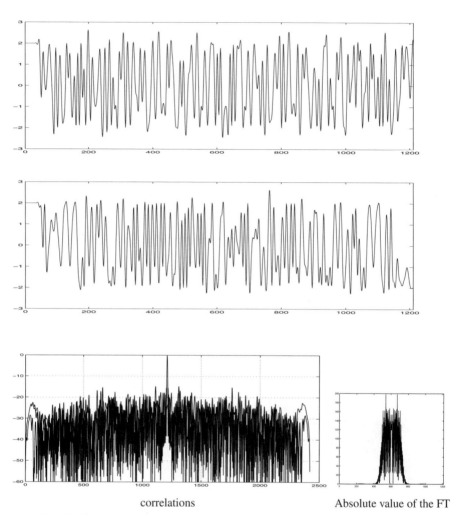

correlations   Absolute value of the FT

Fig. 15.15. Real parts of interpolated sequences and their correlations, $N = 1211$

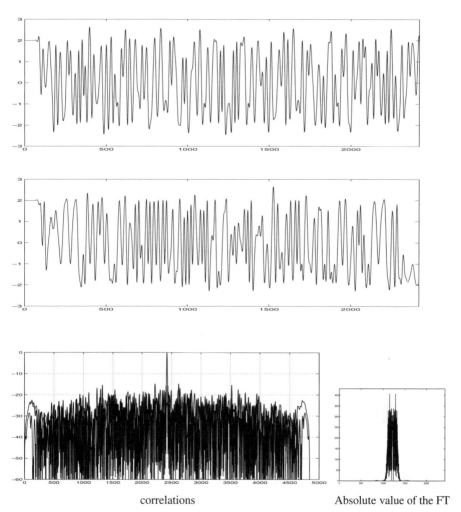

correlations   Absolute value of the FT

Fig. 15.16. Real parts of interpolated sequences and their correlations, $N = 2421$

16

Problems

A key goal of this text is to use known results on linear frequency modulated (FM) chirps and their applications to sonar and radar imaging systems as motivations for systematic sequence design procedures. Zak space (ZS), representation plays a major role in this effort providing a framework for compact and structured representation for echo analysis and for decoupling the two defining characteristics of linear FM chirps, their permutations and modulations. Many mathematically interesting and important topics for applications have only been briefly mentioned, if at all. The following set of problems points the way to potential extensions, generalizations and other applications of the constructions developed in the text.

Variations and Extensions

The design of sequences in Zak space is motivated by the ZS representation of discrete chirps. The modulated permutation sequences have ZS representations

$$[E_\phi \quad \cdots \quad E_\phi] D(\mathbf{x}), \quad \phi \in Perm(L), \mathbf{x} \in \mathbb{C}^{LR},$$

where the permutation matrix E_ϕ is repeated R times.

16.1. Is it possible to construct good correlation sequences by replacing R copies of the permutation matrix E_ϕ by R distinct permutation matrices having some relationship to each other?

$$[E_{\phi_1} \quad \cdots \quad E_{\phi_R}] D(\mathbf{x}), \quad \phi_r \in Perm(L), \mathbf{x} \in \mathbb{C}^{LR}.$$

16.2. In Section 15.2, we constructed real-valued sequences from

$$\begin{bmatrix} E_\phi \\ E_{\phi\rho} \end{bmatrix}$$

where ρ is the time reversal. Are there other ways to construct sequences with good correlation and properties by vertically stacking permutation matrices, either all the same or different?

M. An et al., *Ideal Sequence Design in Time-Frequency Space*,
DOI 10.1007/978-0-8176-4738-4_16,
© Birkhäuser Boston, a part of Springer Science+Business Media, LLC 2009

$$\begin{bmatrix} E_{\phi_1} \\ \vdots \\ E_{\phi_R} \end{bmatrix} D(\mathbf{x}), \quad \phi \in Perm(L), \ \mathbf{x} \in \mathbb{C}^L.$$

16.3. Is it possible to construct good correlation sequences by replacing permutation matrices by repeated partial permutation matrices? Write

$$E_\phi = [E_0 \ E_1], \quad \phi \in Perm(L).$$

What are the properties of sequences whose ZS representations have the form

$$[E_0 \ E_0 \ \cdots \ E_0]$$

or variations on this? Some nonunit discrete chirps have ZS representations of this form.

Echo Analysis

In Chapter 14 we developed methods for echo analysis of zero-padded discrete chirps and modulated permutation sequences based on ZS windows. An algorithm was discussed which extracted echo information by image processing.

16.4. Can other windows be used to extract echo information adapted to application?

16.5. Can information over complete $K \times L$ Zak space be used to extract echo information and remove the ambiguity inherent in the windowing approach?

16.6. Are there noise statistics for which image processing in Zak space is better than matched filtering for extracting echo parameters?

16.7. Varying the factorizations of N leads to different time-frequency representations of sequences and echoes. Can different factorizations lead to processing algorithms which extract more information from echoes?

The ∗-Permutation Condition

16.8. The ∗-permutation condition which is available only for odd integer L leads to modulated permutation sequences satisfying pairwise ideal correlation.

Is there a systematic procedure for relaxing this condition, in the sense described in the text for trading off ideal correlation, to construct sequence sets having predictable correlation properties?

16.9. In the text, methods were introduced for constructing unit ∗-permutations. Three methods were introduced for constructing additional ∗-permutations from known ∗-permutations: shifting, inversion and the tensor product construction.

Are there more rules for constructing ∗-permutations than appeared in this text and can these rules be used for constructing sequence sets satisfying pairwise ideal correlations?

Development of Acyclic Theory

Zero-padding permits acyclic convolution and correlation to be computed by larger size cyclic convolution and correlation. For sufficiently large size problems the availability of the Cooley–Tukey fast Fourier transform made this approach the most computationally efficient method. Small size acyclic computations are usually made directly by specialized programming based on sophisticated mathematics. Algorithms justifying these specialized programs were developed by S. Winograd [40] among others and often involved number theoretic concepts. The history of this development is especially rich, combining both software and hardware advancements.

We come now to the problem of waveform design. The primary focus of sequence design in this text is on the construction of periodic sequences. The periodic properties of these sequences, ideal correlation in particular, are the same. It is a common mythology that good periodic correlation usually implies good acyclic correlation. The actual case is that the acyclic correlation properties of the sequences constructed in this text can vary significantly. As a result, the ambiguity properties can also be very significantly affected in practice by the choice of sequence sets.

Periodic signal design is highly developed because the periodic mathematics on which this design is based is highly developed and more commonly understood.

16.10. Develop concepts for waveform and sequence design for good acyclic correlation by the methods introduced for echo analysis of zero-padded sequences.

16.11. Develop concepts for waveform and sequence design for good acyclic correlation by the number theoretic methods used for small size Fourier transform, cyclic convolution and cyclic correlation. The results of this design approach will probably involve special sizes and data rearrangement.

Multidimensional Signal Representations

Are there other "Zak transforms" that can be used in sequence design strategies? Higher dimensional Zak transforms can be used for over-sampled discrete chirps. A more mathematical problem is based on the fact that the Zak transform can be interpreted as an intertwining operator for two representations of the finite Heisenberg group. An intertwining operator of other finite groups can be used to some advantage in applications.

For example, in [2] an intertwining operator is defined for the affine group over $\mathbb{Z}/N$. Results are closely related to the concept of multiplicative characters [33, 40]. The dimension of the representation varies with the primary factorization of N. Work is under way to determine the natural representation of signals to see if this representation can be used to image Doppler effects on signals. As above, the framework can be turned around to serve as a sequence design tool.

In [33] a part of multiplicative character theory is used (primitive multiplicative characters) to construct periodic sequences having relatively ideal autocorrelation properties. These sequences have many zeros. Work is under way to see if the affine group intertwining operator can significantly extend these results.

Binary Sequences

The procedures in this text depend upon the assumption that signal separation is implemented by correlation defined in terms of standard shifts. For echo analysis, as well as several other applications, standard shift is inherently part of the physical process. The finite Zak transform is defined with this in mind.

For some applications standard shift plays no necessary role, but signal separation is still a major problem. We begin by defining new correlations and Zak transform methods matched to these correlations and set as a goal the development of systematic design procedures which lead to sequences having good correlation properties with respect to these new correlations.

Suppose $R > 1$ is an integer and $L = 2^R$. Set

$$J = (\mathbb{Z}/2)^R.$$

An element $\mathbf{n} \in J$ is written

$$\mathbf{n} = (n_0, \quad \ldots, \quad n_{R-1}), \quad n_r \in \mathbb{Z}/2.$$

The sum of two elements $\mathbf{n}$ and $\mathbf{m}$ in J is defined by

$$\mathbf{n} + \mathbf{m} = (n_0 + m_0, \quad \ldots, \quad n_{R-1} + m_{R-1}),$$

where $n_r + m_r$ is taken in $\mathbb{Z}/2$. For $\mathbf{n} \in J$ define $n \in \mathbb{Z}/L$, $L = 2^R$, by

$$n = n_0 + \quad \cdots \quad + n_{R-1}2^{R-1}.$$

J is ordered by the rule that $\mathbf{n}$ is the n-th element in J.

Set

$$S = \begin{bmatrix} 0 & 1 \\ 1 & 0 \end{bmatrix}.$$

For $\mathbf{m} \in J$ define

$$\mathcal{S}(\mathbf{m}) = S(m_{R-1}) \otimes \cdots \otimes S(m_0).$$

$\mathcal{S}(\mathbf{m})$ is the $L \times L$ binary shift matrix.

For $\mathbf{x}$ and $\mathbf{y} \in \mathbb{C}^L$ define the *binary* convolution $\mathbf{x} * \mathbf{y}$ in $\mathbb{C}^L$ by

$$\mathbf{x} * \mathbf{y} = \sum_{\mathbf{n} \in J} x_{\mathbf{n}} \mathcal{S}(\mathbf{n}) \mathbf{y}.$$

Since $\mathbf{m} + \mathbf{n} = \mathbf{m} - \mathbf{n}$, binary cyclic correlation $\mathbf{x} \circ \mathbf{y}$ in $\mathbb{C}^L$ is defined by

$$\mathbf{x} \circ \mathbf{y} = \mathbf{x} * \mathbf{y}^*.$$

Set

$$F = \begin{bmatrix} 1 & 1 \\ 1 & -1 \end{bmatrix}.$$

Suppose R is an even integer, $R = 2R_1$ and $L_1 = 2^{R_1}$. F_{R_1} is the $L_1 \times L_1$ matrix formed by the tensor product of R_1 copies of F. For $\mathbf{x} \in \mathbb{C}^L$, the *binary* Zak transform of $\mathbf{x}$ is defined by first computing

$$(I_{L_1} \otimes F_{R_1}) P(L, L_1) \mathbf{x} = [X(n)]_{0 \le n < L_1}, \qquad X(n) \in \mathbb{C}^{L_1}$$

and then setting

$$Z\mathbf{x} = [X(0) \quad \cdots \quad X(L_1 - 1)].$$

16.12. Use the preceding framework to develop a signal design strategy for binary sequences, i.e., sequences with ± 1, having good correlation properties.

16.13. Extend the above framework to shifts

$$S_3 = \begin{bmatrix} 0 & 0 & 1 \\ 1 & 0 & 0 \\ 0 & 1 & 0 \end{bmatrix},$$

and larger size shift matrices and develop a signal design strategy.

Finite Fields

Finite field theory has played no role in this text except in several examples. In this way our methods and results are complementary to the extensive theory of finite field signal design appearing, for example, in the fundamental text *Signal Design for Good Correlation* by S.W. Golomb and G. Gong [20] and the chapter "Sequences with low correlation," in *Handbook of Coding Theory* by T. Helleseth and P.V. Kumar [23]. Many additional works are referenced in these works.

An N-point finite Fourier transform can be defined over any finite field containing N-th roots of unity [5, 20]. In the following example we consider a concrete approach which has the advantage of realizing correlation in terms of complex conjugation. Take $N \equiv 3 \bmod 4$ and consider $\mathbb{Z}[i]/N$, the Gaussian integers modulo N.

Example 16.1 For $N = 19$, $r = 7 + 3i$, we have

$$r^5 = 1 \quad \text{and} \quad r^{-1} = r^*.$$

The 5-point Fourier transform matrix over $\mathbb{Z}[i]/19$ is

$$F(5)_{19} = \left[r^{jk} \right]_{0 \le j, \, k < 5} = \begin{bmatrix} 1 & 1 & 1 & 1 & 1 \\ 1 & 7 + 16i & 2 + 15i & 2 + 4i & 7 + 3i \\ 1 & 2 + 15i & 7 + 3i & 7 + 16i & 2 + 4i \\ 1 & 2 + 4i & 7 + 16i & 7 + 3i & 2 + 15i \\ 1 & 7 + 3i & 2 + 4i & 2 + 15i & 7 + 16i \end{bmatrix}.$$

We have

$$F(5)_{19}^2 = 5R_5, \quad F(5)_{19}^4 = 6I_5 = 5^2 I_5.$$

$F(5)_{19}$ diagonalizes S_5. We can define the Zak transform over $\mathbb{Z}[i]/19$ and construct permutation sequences satisfying ideal autocorrelation.

Example 16.2 For $N = 7$, $r = 2 + 2i$, we have

$$r^8 = 1 \quad \text{and} \quad r^{-1} = r^*.$$

The 8-point Fourier transform matrix over $\mathbb{Z}[i]/19$ is

$$F(8)_7 = \left[r^{jk}\right]_{0 \le j,\, k < 8} = \begin{bmatrix} 1 & 1 & 1 & 1 & 1 & 1 & 1 & 1 \\ 1 & 2+5i & 6i & 5+5i & 6 & 5+2i & i & 2+2i \\ 1 & 6i & 6 & i & 1 & 6i & 6 & i \\ 1 & 5+5i & i & 2+5i & 6 & 2+2i & 6i & 5+2i \\ 1 & 6 & 1 & 6 & 1 & 6 & 1 & 6 \\ 1 & 5+2i & 6i & 2+2i & 6 & 2+5i & i & 5+5i \\ 1 & i & 6 & 6i & 1 & i & 6 & 6i \\ 1 & 2+2i & i & 5+2i & 6 & 5+5i & 6i & 2+5i \end{bmatrix}.$$

$F(8)_7$ has all the desired properties.

Ambiguity Functions

The ambiguity function and ambiguity surface of a signal are important design criteria in radar and sonar applications, where both the target position and target velocity must be simultaneously determined. Ideally, the ambiguity function of a signal should be a thumbtack. A discussion of this topic can be found in R.E. Blahut's chapter in [6].

Generally the ambiguity function of a chirp is not good. From a Zak space point of view, translations in time cannot be distinguished from translations in frequency. We have found that the ambiguity function of some permutation sequence is superior to that of the same size chirp. Is this a result of a higher number of parameters describing the permutation sequence as compared with the chirp and is this phenomena related to the randomness of the permutation sequence? Is there any relationship to random-ness in communication theory? Specifically, is there a randomness measurement, for example described in [20], which can be computed for permutation sequences tying together good ambiguity functions with covert signals in spread spectrum communi-cation systems?

References

1. W.O. Alltop, "Complex sequences with low periodic correlations," *IEEE Trans. Info. Theo.*, **26**(3), 350–354, 1980.
2. M. An and R. Tolimieri, "Affine group Zak transform," in preparation.
3. J.J. Benedetto and J. Donatelli, "Ambiguity function and frame theoretic properties of periodic zero autocorrelation functions," *IEEE J. of Selected Topics in Signal Processing*, **1**, 2007, 6–20.
4. B.C. Berndt, R.J. Evans and K.S. Williams, *Gauss and Jacobi Sums*, Wiley Interscience, New York, 1998.
5. R.E. Blahut, *Theory and Practice of Error Control Codes*, Addison-Wesley, Reading, MA, 1983.
6. R.E. Blahut, "Theory of remote surveillance algorithms, radar and sonar," Part I, *The IMA Volumes in Mathematics and Its Applications*, **32**, A. Friedman and W. Miller Jr., Eds., 1991.
7. A.K. Brodzik, "On the Fourier transform of finite chirps," *IEEE Signal Processing Letters*, **13**(9), 541–544, 2006.
8. A.K. Brodzik, "Characterization of Zak space support of the finite chirp," *IEEE Trans. Info. Theo.*, **53**(6), 2007.
9. A.K. Brodzik and R. Tolimieri, "Bat chirps with good properties: Zak space construction of perfect polyphase sequences," submitted to *IEEE Trans. Info. Theory*.
10. P.G. Casazza and M. Fickus, "Chirps on finite cyclic groups," *Proc. SPIE 2005*.
11. D.C. Chu, "Polyphase codes with good periodic correlation properties," *IEEE Trans. Info. Theo.*, **IT**(18), 531–532, 1972.
12. H. Chung and P.V. Kumar, "A new general construction for generalized bent functions," *IEEE Trans. Info. Theo.*, **IT**(35), 206–209, 1989.
13. J.W. Cooley and J.W. Tukey, "An algorithm for machine calculation of complex Fourier series," *Math. Comp.*, **19**, 297–301, 1965.
14. H.G. Feichtinger and T. Strohmer, Eds. *Gabor Analysis and Algorithms*, Applied Numerical Harmonic Analysis Series, Birkhäuser, Cambridge, 1997.
15. R. Frank and S. Zadoff, "Phase shift pulse codes with good periodic correlation properties," *IEEE Trans. Info. Theo.*, **IT**(19), 244, 1973.

16. J. Gabardo, "Some problems related to the distributional Zak transform," *Harmonic analysis and applications. In Honor of John J. Benedetto. Applied and Numerical Harmonic Analysis*, 101–126, 2006, Christopher Heil, Ed., Birkhäuser.

17. D. Gabor, "Theory of communications," *JIEE*, **93**, 429–459, 1946.

18. R.A. Games, "Cross correlation of m-sequences and GMW-sequences with the same primitive polynomial," *Discrete Appl. Math.*, **12**, 139–146, 1985.

19. C.F. Gauss *Arithmetische Untersuchungen*. Chelsea, New York, 1965.

20. S.W. Golomb and G. Gong, *Signal Design for Good Correlation*, Cambridge University Press, New York, 2005.

21. G. Gong, "Theory and application of q-ary interleaved sequences," *IEEE Trans. Info. Theo.*, **IT**(41), 400–411, 1995.

22. R.C. Heimiller, "Phase shift pulse codes with good periodic correlation properties," *IRE Trans. Info. Theo.*, **IT**(7), 254–257, 1961.

23. T. Helleseth and P.V. Kumar, "Sequences with Low Correlation," *Handbook of Coding Theory*, Chapter 21, V.S. Pless and W.C. Huffman, Eds., Elsevier Science B.V., 1998.

24. K. Ireland and M. Rosen, *A Classical Introduction to Modern Number Theory*, Springer-Verlag, New York, 1982.

25. A.J.E.M. Janssen, "The Zak transform: A signal transform for sampled time-continuous signals," *Phillips Journal of Research*, **43**, 23–69, 1988.

26. J.R. Klauder, A.C. Price, S. Darlington and W.J. Albersheim, "The theory and design of chirp radars," *Bell System Tech. J.*, **39**, 745–808, 1960.

27. P.V. Kumar, R.A. Scholtz and L.R. Welch, "Generalized bent functions and their properties," *J. Combin. Theo.*, **A**(40), 90–107, 1985.

28. R.M. Lerner, "Representation of signals," *Lectures on Communication System Theory*, E.J. Baghdady, Ed., 203–242, McGraw-Hill, New York, 1961.

29. B. Mahafza and A. Elsherbeni, *MATLAB Simulations for Radar Systems Design*, Chapman & Hall/CRC, New York, 2004.

30. S. Pei and M. Yeh, "Time and frequency split Zak transform for finite Gabor expansion," *Signal Process*, **52**(3), 323–341, 1996.

31. B.M. Popovic, "Generalized chirp-like polyphase sequences with optimum correlation properties," *IEEE Trans. Info. Theo.*, **IT**(38), 1406–1409, 1992.

32. D.V. Sarwate, "Bounds on cross correlation and autocorrelation of sequences," *IEEE Trans. Info. Theo.*, **IT**(25), 720–724, 1979.

33. R.A. Scholtz and L.R. Welch, "Group characters: Sequences with good correlation properties," *IEEE Trans. Info. Theo.*, **IT 24**(5), 1978.

34. C.L. Siegel, "Lectures on the analytical theory of quadratic forms," *Institute for Advanced Study Notes*, 1934.

35. T. Strohmer and R. Heath, "Grassmannian frames with applications to coding and communications," *Appl. Comp. Harm. Anal.*, **14**(3), 257–275, 2003.

36. N. Suehiro and M. Hatori, "Modulatable orthogonal sequences and their applications to SSMA systems," *IEEE Trans. Info. Theo.*, **IT**(34), 93–100, 1988.

37. R. Tolimieri, "Multispectral SAR Imaging for Material Identification," Internal published technical report, AFRL/MLKH, 2003.

38. R. Tolimieri and M. An, *Time-frequency Representations*, Birkhäuser, Boston, 1998.

39. R. Tolimieri and M. An, "MISAR: Waveform Design and Filtering in Zak Space," Progress report, Brooks City Base, 2004.

40. R. Tolimieri, M. An and C. Lu, *Algorithms for Discrete Fourier Transform and Convolution*, 2nd Ed., Springer-Verlag, New York, 1997.

41. R. Tolimieri, M. An and C. Lu, *Mathematics of Multidimensional Fourier Transform Algorithms*, 2nd Ed., Springer-Verlag, New York, 1997.

42. A. Weil, "Sur certains groups d'operateurs unitaires," *Acta Math.*, **111**, 143–211, 1964.

43. J. Zak, "Finite translations in solid state physics," *Phys. Rev. Lett.*, **19**, 1385–1397.

Index

Applied and Numerical Harmonic Analysis

J.M. Cooper: *Introduction to Partial Differential Equations with MATLAB* (ISBN 978-0-8176-3967-9)

C.E. D'Attellis and E.M. Fernández-Berdaguer: *Wavelet Theory and Harmonic Analysis in Applied Sciences* (ISBN 978-0-8176-3953-2)

H.G. Feichtinger and T. Strohmer: *Gabor Analysis and Algorithms* (ISBN 978-0-8176-3959-4)

T.M. Peters, J.H.T. Bates, G.B. Pike, P. Munger, and J.C. Williams: *The Fourier Transform in Biomedical Engineering* (ISBN 978-0-8176-3941-9)

A.I. Saichev and W.A. Woyczyński: *Distributions in the Physical and Engineering Sciences* (ISBN 978-0-8176-3924-2)

R. Tolimieri and M. An: *Time-Frequency Representations* (ISBN 978-0-8176-3918-1)

G.T. Herman: *Geometry of Digital Spaces* (ISBN 978-0-8176-3897-9)

A. Procházka, J. Uhlíř, P.J.W. Rayner, and N.G. Kingsbury: *Signal Analysis and Prediction* (ISBN 978-0-8176-4042-2)

J. Ramanathan: *Methods of Applied Fourier Analysis* (ISBN 978-0-8176-3963-1)

A. Teolis: *Computational Signal Processing with Wavelets* (ISBN 978-0-8176-3909-9)

W.O. Bray and Č.V. Stanojević: *Analysis of Divergence* (ISBN 978-0-8176-4058-3)

G.T Herman and A. Kuba: *Discrete Tomography* (ISBN 978-0-8176-4101-6)

J.J. Benedetto and P.J.S.G. Ferreira: *Modern Sampling Theory* (ISBN 978-0-8176-4023-1)

A. Abbate, C.M. DeCusatis, and P.K. Das: *Wavelets and Subbands* (ISBN 978-0-8176-4136-8)

L. Debnath: *Wavelet Transforms and Time-Frequency Signal Analysis* (ISBN 978-0-8176-4104-7)

K. Gröchenig: *Foundations of Time-Frequency Analysis* (ISBN 978-0-8176-4022-4)

D.F. Walnut: *An Introduction to Wavelet Analysis* (ISBN 978-0-8176-3962-4)

O. Bratteli and P.E.T. Jorgensen: *Wavelets through a Looking Glass* (ISBN 978-0-8176-4280-8)

H.G. Feichtinger and T. Strohmer: *Advances in Gabor Analysis* (ISBN 978-0-8176-4239-6)

O. Christensen: *An Introduction to Frames and Riesz Bases* (ISBN 978-0-8176-4295-2)

L. Debnath: *Wavelets and Signal Processing* (ISBN 978-0-8176-4235-8)

J. Davis: *Methods of Applied Mathematics with a MATLAB Overview* (ISBN 978-0-8176-4331-7)

G. Bi and Y. Zeng: *Transforms and Fast Algorithms for Signal Analysis and Representations* (ISBN 978-0-8176-4279-2)

J.J. Benedetto and A. Zayed: *Sampling, Wavelets, and Tomography* (ISBN 978-0-8176-4304-1)

E. Prestini: *The Evolution of Applied Harmonic Analysis* (ISBN 978-0-8176-4125-2)

O. Christensen and K.L. Christensen: *Approximation Theory* (ISBN 978-0-8176-3600-5)

L. Brandolini, L. Colzani, A. Iosevich, and G. Travaglini: *Fourier Analysis and Convexity* (ISBN 978-0-8176-3263-2)

W. Freeden and V. Michel: *Multiscale Potential Theory* (ISBN 978-0-8176-4105-4)

O. Calin and D.-C. Chang: *Geometric Mechanics on Riemannian Manifolds* (ISBN 978-0-8176-4354-6)

Applied and Numerical Harmonic Analysis (Cont'd)

J.A. Hogan and J.D. Lakey: *Time-Frequency and Time-Scale Methods*
(ISBN 978-0-8176-4276-1)

C. Heil: *Harmonic Analysis and Applications* (ISBN 978-0-8176-3778-1)

K. Borre, D.M. Akos, N. Bertelsen, P. Rinder, and S.H. Jensen: *A Software-Defined GPS and Galileo Receiver* (ISBN 978-0-8176-4390-4)

T. Qian, V. Mang I, and Y. Xu: *Wavelet Analysis and Applications* (ISBN 978-3-7643-7777-9)

G.T. Herman and A. Kuba: *Advances in Discrete Tomography and Its Applications*
(ISBN 978-0-8176-3614-2)

M.C. Fu, R.A. Jarrow, J.-Y. J. Yen, and R.J. Elliott: *Advances in Mathematical Finance*
(ISBN 978-0-8176-4544-1)

O. Christensen: *Frames and Bases* (ISBN 978-0-8176-4677-6)

P.E.T. Jorgensen, K.D. Merrill, and J.A. Packer: *Representations, Wavelets, and Frames*
(ISBN 978-0-8176-4682-0)

M. An, A.K. Brodzik, and R. Tolimieri: *Ideal Sequence Design in Time-Frequency Space*
(ISBN 978-0-8176-4737-7)

Printed in the United States of America